Nanotechnology
and the
Environment

Nanotechnology
and the
Environment

Kathleen Sellers
Christopher Mackay
Lynn L. Bergeson
Stephen R. Clough
Marilyn Hoyt
Julie Chen
Kim Henry
Jane Hamblen

CRC Press
Taylor & Francis Group
Boca Raton London New York

CRC Press is an imprint of the
Taylor & Francis Group, an **informa** business

CRC Press
Taylor & Francis Group
6000 Broken Sound Parkway NW, Suite 300
Boca Raton, FL 33487-2742

First issued in paperback 2019

© 2009 by Taylor & Francis Group, LLC
CRC Press is an imprint of Taylor & Francis Group, an Informa business

No claim to original U.S. Government works

ISBN-13: 978-1-4200-6019-5 (hbk)
ISBN-13: 978-0-367-38706-8 (pbk)

Library of Congress Cataloging-in-Publication Data

Nanotechnology and the environment / authors, Kathleen Sellers ... [et al.].
    p. cm.
  Includes bibliographical references and index.
  ISBN 978-1-4200-6019-5 (alk. paper)
  1. Nanotechnology--Environmental aspects. I. Sellers, Kathleen. II. Title.

T174.7.N373193 2008
620'.5--dc22
                                                2008019075

Visit the Taylor & Francis Web site at
http://www.taylorandfrancis.com

and the CRC Press Web site at
http://www.crcpress.com

Contents

Acknowledgments

The authors gratefully acknowledge the scientists and engineers who reviewed drafts of this manuscript and provided valuable perspective, including:

Larry S. Andrews, Ph.D., Rohm & Haas
Janis Bunting, General Dynamics
Stefanie Giese-Bogdan, Dr.rer.nat., 3M
Laurie Gneiding, AMEC
Martin P. A. Griffin, Wisconsin Department of Natural Resources
Kimberly A. Groff, Ph.D., AMEC
Richard Johnson, Ph.D., Oregon Health and Science University
Maria Victoria Peeler, Washington State Department of Ecology
Nora Savage, Ph.D., U.S. Environmental Protection Agency
John Schupner, AMEC
Jack Spadaro, Ph.D., AMEC

The authors appreciate the assistance of the following contributors:

Robin Weinbeck, AMEC
Beth Auerbach, Bergeson & Campbell, P.C.
Ira Dassa, Bergeson & Campbell, P.C.
Elizabeth Algeo, AMEC
Kevin Haines, AMEC
Denise Ladebauche, AMEC
Elizabeth Martin, AMEC

Finally, the AMEC team gratefully acknowledges the support for this project under the AMEC Technical Council's Research and Development Program.

About the Authors

Kathleen Sellers, PE, has worked on developing solutions to a variety of environmental problems during her 20 years of experience. She is particularly intrigued with exploring emerging environmental issues and their solutions. An environmental engineer and chemist, Sellers' work has ranged from analysis and bioassay of environmental samples through comprehensive field characterizations, and extended to treatment process development, permitting, and negotiation of regulatory solutions. Sellers published an engineering textbook entitled *Fundamentals of Hazardous Waste Site Remediation* (CRC Press/Lewis Publishers, 1999), then edited and co-authored *Technical Brief: Endocrine Disrupting Compounds and Implications for Wastewater Treatment* (Water Environment Research Foundation, 2005) and *Perchlorate: Environmental Problems and Solutions* (Taylor & Francis / CRC Press, 2006) before editing and co-writing this book.

Chris E. Mackay, PhD, is an environmental chemist and toxicologist with more than 10 years of experience conducting site, product, and chemical compliance assessments. Mackay has worked extensively at the interface of industrial, pharmaceutical, and agricultural chemistry and environmental biology. To answer challenging environmental and toxicology questions, he has constructed statistical models such as quantitative structure-activity relationships (QSAR) and physiologically based pharmacokinetics (PBPK), as well as first principal models for stochastic competition and thermodynamic energy transfer. These models have been applied to illustrate and predict not only the transport, fate, and impact of chemical and biological stresses to human populations and aquatic and terrestrial wildlife, but also to optimize industrial and agricultural productivity and economy. Along with providing technical guidance in product and patent development, Mackay has been involved in designing chemical and toxicological assessments to support product development and to comply with regulatory requirements under various consumer and environmental protection statutes in both the United States and Europe. He also has considerable experience in regulatory chemistry and toxicology.

Lynn L. Bergeson is Managing Director of Bergeson & Campbell, P.C. (B&C), a Washington, D.C. law firm concentrating on conventional and engineered nanoscale chemical, pesticide, and other specialty chemical product regulation and approval matters, domestic and foreign chemical classification, chemical product litigation, Food and Drug Administration-regulated product approval, and associated business issues. Bergeson is also president of The Acta Group, L.L.C. and The Acta Group EU, Ltd., B&C's consulting affiliates, with offices in Washington, DC and Manchester, UK, respectively. She counsels clients on health, safety, science policy, and related legal and regulatory aspects of traditional domestic chemical regulatory programs under the Federal Insecticide, Fungicide, and Rodenticide Act (FIFRA) and the Toxic Substances Control Act (TSCA), as well as issues pertinent to nanotechnology and other emerging transformative technologies.

Stephen R. Clough, PhD, DABT, graduated from the University of Michigan with an MS in Water Quality (1984) and a PhD in Toxicology (1988). His career experience includes the design and management of mammalian and aquatic toxicology laboratories for the planning and analysis of safety evaluation studies addressing both commercial products and industrial waste streams. He was later certified as a Diplomate by the American Board of Toxicology and has since served more than 20 years as a senior environmental toxicologist. He specializes in benthic bioassessments and the evaluation of point and non-point impacts to the aquatic macro-invertebrate community. Clough has worked at more than 50 hazardous waste sites where he has evaluated the fate, exposure and ecological effects of a wide range of toxicants in various types of media, including heavy metals, pesticides, dioxins/furans, perchlorate, polycyclic aromatic hydrocarbons, and residuals from chemical, pharmaceutical, pulp/paper/forestry, mining, and automotive industries. He has a broad understanding of projects and/or regulations that fall under CERCLA, RCRA, TSCA, FIFRA, and the Clean Water Act, particularly in US EPA Regions I through IV.

Marilyn Hoyt has more than 25 years of experience in materials science and environmental consulting. She has provided chemical measurement support for the research and development of strategic defense systems as part of a multidisciplinary team, as well as managed numerous environmental measurement programs. Hoyt has directed a full service environmental laboratory and has managed studies requiring original research with data collection and interpretation. She is familiar with the full array of characterization and instrumental measurement techniques applicable to nanoparticle measurement, as well as the particular challenges involved in environmental analyses as compared to laboratory studies.

Julie Chen, PhD, is currently one of the three co-directors of the University of Massachusetts Lowell Nanomanufacturing Center. She is responsible for the Nanomanufacturing Center of Excellence (NCOE), a state-funded center with the mission of fundamental scientific and applied, industry-collaborative research on environmentally benign, commercially viable manufacturing with nanoscale control. Chen is also the co-director of the Advanced Composite Materials and Textile Research Laboratory at the University of Massachusetts Lowell, where she is a professor of mechanical engineering. Chen was the program director of the Materials Processing and Manufacturing and the Nanomanufacturing Programs in the Division of Design, Manufacture, and Industrial Innovation at the National Science Foundation from 2002 to 2004. Chen has been on the faculty at Boston University, a NASA-Langley Summer Faculty Fellow, a visiting researcher at the University of Orleans and Ecole Nationale Supérieure d'Arts et Métiers (ENSAM-Paris), and an invited participant in the National Academy of Engineering, Frontiers of Engineering Program (US, 2001, US-Germany, 2005, and Indo-US, 2006).

Chen received her PhD, MS, and BS in Mechanical Engineering from MIT. She has more than 20 years of experience in the mechanical behavior and deformation of fiber structures, fiber assemblies, and composite materials, with an emphasis on composites processing and nanomanufacturing. Examples include analytical modeling and novel experimental approaches to electrospinning and controlled patterning of nanofibers, and nanoheaters, as well as forming, energy absorption, and failure of textile reinforcements for structural (biomedical to automotive) applications.

Kim Henry brings her experience as a hydrogeologist to the fate and transport of nanoparticles in the environment, and the application of those principles to waste-water treatment. She has a BA in geological sciences from Harvard University and an MS in environmental science and engineering from Rice University. Henry has more than 20 years of experience in characterizing and remediating environmental contaminants at a variety of sites, including former manufactured gas plants, rail yards, gasoline stations, bulk storage terminals, chemical refineries, and military facilities. She has negotiated highly controversial projects with regulatory agencies and presented the findings of site investigations at public meetings and press conferences. Henry is intrigued by the concept of science writing, of presenting leading scientific issues accurately and evocatively to the public. She is also the author of four children's books.

Jane Hamblen has more than 28 years of experience in risk assessment, environmental health, and biology. As a senior health scientist, Hamblen is responsible for managing risk assessment projects and for technical assistance on a variety of human health projects. Hamblen's expertise lies in exposure assessment, specializing in both deterministic and probabilistic analyses. Most recently, she has applied those skills to the assessment of nanomaterials and the identification of toxic effects from exposure to nanomaterials. In addition, she has evaluated the public health impacts associated with PCBs, chlorinated organics, mercury, resorcinol, pesticides, gasoline spills, and ash generated by power plants. Hamblen has also developed risk-specific chemical concentrations used to derive clean-up goals, and evaluated public health risks associated with remedial alternatives. She has co-authored several published technical papers, including articles on the use of Monte Carlo methodology in exposure assessment. Hamblen's extensive project experience has included human health risk assessments conducted under CERCLA and RCRA. Her project management experience ranges from baseline risk assessments to dose reconstruction projects. She has managed toxic tort cases that allege health effects from chemical exposure and worked in the area of regulatory compliance for worker and community right-to-know legislation.

1 Introduction

Kathleen Sellers
ARCADIS U.S., Inc.

CONTENTS

In early 2007, the United Nations reported that nanotechnology, which then accounted for approximately 0.1% of the global manufacturing economy, would grow to 14% of the market by 2014. This market share would correspond to $2.6 trillion in U.S. dollars [1]. What accounts for this explosive growth? And what does it mean for our environment? This book provides perspective on those questions based on the current state of the science.

Nanotechnology is a field of applied science concerned with the control of matter at dimensions of roughly 1 to 100 nanometers (nm) [1]. (1 nm is one-billionth of a meter.) At the particle size of 1 to 100 nm, nanoscale materials may have different molecular organizations and properties than the same chemical substances in a larger size. Nano-sized chemicals can have different properties due to [2]:

- Increased relative surface area per unit mass, which can increase physical strength and chemical reactivity
- In some cases, the dominance of quantum effects at the nanometer size, which changes basic material properties

These unique properties offer revolutionary means to optimize a variety of products, including electronics, textiles, paintings and coatings, pharmaceuticals, and personal care products. And these unique properties mean that nanoscale materials can behave differently in the human body and the environment than the corresponding macro-scale materials.

Similarly, revolutionary developments during the past two centuries offer cautionary tales. In the 1800s, gaslights illuminated the Industrial Revolution. Engineers had devised ways to manufacture gas from the pyrolysis of coal or oil. A hundred years later, the residuals of that process stained soils bright blue with cyanide compounds and contaminated groundwater with tar residuals. Those historic manufactured gas plants had come to represent hazardous waste rather than progress. Developments in the 1900s provide a further example. In 1979, suppliers began adding methyl-tertiary-butyl ether (MTBE) to gasoline in the United States to replace lead as an octane enhancer. Later, adding MTBE to gasoline fulfilled the oxygenate require-ments in the 1990 Clean Air Act Amendments intended to reduce smog production. The use of MTBE, however, created another set of environmental problems. Liq-uid and vapor leaks from underground storage tanks have led to widespread MTBE contamination in groundwater. The U.S. Geological Survey surveyed water quality in nearly all 50 states in the 1990s. Of the 4023 groundwater samples collected, 10% contained detectable MTBE at an average concentration of 280 micrograms per liter (µg/L) [3], well above the U.S. Environmental Protection Agency's Health Advisory of 20 to 40 µg/L in drinking water [4]. These examples illustrate the unintended consequences that can result from rapid industrial progress.

1.1 POTENTIAL REWARDS

Nanotechnology offers the potential for tantalizing rewards. Amid the hyperbole and hype, many experts believe that nanotechnology may offer substantial advan-tages. Consider the following examples [2, 5, 6]:

- *Energy savings.* The U.S. Environmental Protection Agency (U.S. EPA) has cited one estimate that the use of nanotechnology could reduce the energy consumption in the U.S. by more than 14% [5]. For example, the use of nanotechnology-based materials such as lightweight composites and thinner paint coatings can reduce the weight of airplanes and automobiles, and thus their fuel usage. Solid-state lighting may use energy more effi-ciently than conventional lighting. Fuel additives, such as cerium oxide, may increase diesel fuel efficiency.
- *Alternative energy supplies.* Nanotechnology offers the potential to decrease the cost of producing solar cells to enable more widespread use of solar power. Advances in battery manufacturing using nanotechnology may allow for more widespread use of electric vehicles. Finally, with respect to hydrogen fuel cells, nanotechnology can provide more efficient fuel storage methods and improve efficiency.
- *Efficient use of raw materials.* Nanostructured catalysts may decrease the mass of catalysts, particularly platinum, used in some applications. The use of highly effective nano-sized catalysts also can increase production and decrease waste generation. Nanoscale zeolite catalysts, for example, are used now in petroleum cracking. Some nanomaterials may provide sub-stitutes for toxic materials; for example, nanotechnology-based solders can replace lead-based solders.
- *Environmental protection.* Engineers use nanomaterials in wastewater treatment and environmental remediation, as described later in this book. Researchers

also are studying the use of nanotechnology to treat air pollution. Finally, sensors based on nanotechnology can detect some chemical contaminants.

- *Agricultural applications.* Increased biological efficiency could diminish the amount of pesticides being applied. Similarly, nanodevices used for "smart" treatment delivery systems hold promise. Smart field systems detect, locate, and report/apply, as needed, pesticides and fertilizers prior to the onset of symptoms. Nanoparticle delivery systems, including nanocapsules, nanocontainers, and nanocages, could replace conventional emulsifiable concentrates, thus reducing organic solvent content in agricultural formulations, and enhancing dispersity, wettability, and the penetration strength of the droplets. Enhanced use of smart systems also could diminish runoff and avert unwanted movement of pesticides.
- *Medical breakthroughs.* Nanotechnology is used to create artificial bone and may be used in other prosthetic devices in the future. Researchers are studying ways to use nanomaterials in medical imaging and for targeted drug delivery. Probes based on nanotechnology can detect and monitor changes within cells without destroying them.

This range of potential benefits illustrates why so many are excited about the promise of nanotechnology and its potential economic importance.

1.2 POSSIBLE RISKS AND PUBLIC CONCERNS

The nanotechnology revolution also can present risks. The words of some of the stakeholders, beginning with a prominent proponent of nanotechnology, illustrate a range of views regarding potential risks.

- "The state of knowledge with respect to the actual risks of nanotechnology is incomplete." — *United States President's Council of Advisors on Science and Technology,* 2005 [7].
- "It may be that in most cases nanomaterials will not be of human health or ecological concern. However, at this point not enough information exists to assess environmental exposure for most engineered nanomaterials." — *U.S. EPA,* 2007 [8].
- "At present, the toxicological and ecotoxicological risks linked to this expanding technology … cannot be assessed yet. Nanotechnology is moving increasingly into the center of public attention. However, currently it is not yet linked to any great degree to concerns about health and the environment. Over the next few years this could change if the media increasingly will point at components linked with nanotechnology that are harmful to health or the environment (cf. also public debate on genetically modified organisms (GMOs))." — *Federal Institute for Occupational Safety and*

Health (BAuA), Federal Institute for Risk Assessment (BfR), Federal Environment Agency (UBA), Germany, 2006 [9].

- "Knowing the basics about the dangers of new materials is a prerequisite for effective environmental responsibility. With cause for concern, and with the precautionary principle applied, these materials should be considered hazardous until shown otherwise." — *Greenpeace Environmental Trust, 2003* [10].

As nanotechnology penetrates the marketplace and receives some press coverage (Figure 1.1* [11, 12]), the public is just beginning to develop opinions about nanotechnology. Those opinions will shape the market for consumer goods containing nanotechnology. They may affect the concerns of workers who manufacture or incorporate nanomaterials into commercial products. Finally, public opinion influences the development of regulations. Given the possible weight and effects of public opinions, they are worth examining in this Introduction.

Public surveys between 2003 and 2007 showed some common themes and some different perceptions regarding nanotechnology around the world. Table 1.1 shows the results of six surveys of people in the U.S., Great Britain, and Japan between 2003 and 2007. This compilation is not a meta-survey or statistical analysis of the aggregated data, but a summary of some of the key findings of these six surveys.

In general, many people knew little to nothing about nanotechnology. The level of knowledge varied: between approximately 20 and 46% of respondents to various U.S. surveys had some familiarity with nanotechnology, as did 26% of survey respondents in Great Britain and 55% in Japan. Many survey respondents distrusted the ability of the government or corporations to manage nanotechnology wisely. Confidence in the benefits of nanotechnology generally increased with knowledge.

In seeming contrast to the public's relatively limited knowledge of nanotechnology, between mid-2006 and mid-2007 the number of commercially available products containing nanotechnology more than doubled [13]. The prefix "nano" or key words such as "micronized" featured prominently in some marketing materials, suggesting that consumers may favor products with the cachet of nanotechnology — or at any rate, that advertisers believe they will.

In the early years of the nanotechnology revolution, then, regulators and non-governmental organizations (NGOs) acknowledge the lack of information on the risks of nanotechnology. Surveys have shown that much of the public is unfamiliar with nanotechnology. Despite this limited knowledge, survey respondents believed that nanotechnology offers potential risks and rewards. Absent specific knowledge,

---

* Figure 1.1 shows the results of two literature searches that illustrate the increase in the availability of information regarding nanotechnology. A search of the NewsBank database [11] represents information available to the general public. NewsBank incorporates information from 27 news magazines, ranging from People weekly to Popular Science. Figure 1.1 shows the results of a keyword search of NewsBank on "nanotechnology" performed on June 30, 2007. A similar search performed in Academic ASAP™ on InfoTrac Web [12] shows the increase in technical publications. The Academic ASAP database includes more than 14 million articles from academic journals (e.g., Journal of the American Chemical Society), magazines (e.g., Chemical Week), and news (e.g., Pesticide and Chemical News). Values for 2007 in Figure 1.1 were estimated by doubling the numbers of articles published between January 1 and June 30, 2007.

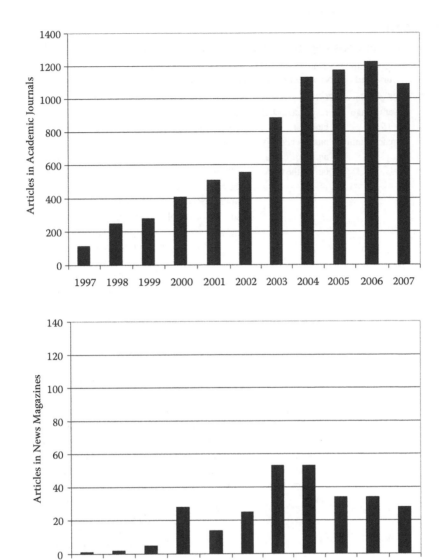

FIGURE 1.1 Articles in news magazines and academic journals regarding "nanotechnology," 1997–2007 [11, 12].

people can speculate about doomsday scenarios. Once provided with information, citizen groups in one recent study believed strongly that the public should be provided more information and heard in governmental decision making. Risk communication by government agencies, NGOs, and corporations engaged in nanotechnology, will help shape the public's opinions as they evolve. Those opinions, in turn, will influence the marketplace and regulatory developments. For now, public concerns over

TABLE 1.1
Public Perceptions Regarding Nanotechnology Based on Opinion Polls

Year	Population Surveyed	Knowledge of Nanotechnology	Perception of Risk	Other Findings	Ref.
2003	1005 people over the age of 15 in Great Britain	26% of respondents had heard of nanotechnology and 19% could define the term.	68% of respondents believed nanotechnology would improve life in the future; 13% believed that the consequences would depend on how nanotechnology was used.	In subsequent workshops with 50 participants in urban areas, researchers further explored perceptions. Participants perceived the benefits might include medical breakthroughs and other enhancements to the quality of life, and hoped for unforeseeable benefits. Concerns included social justice, financial implications, long-term side effects, whether nanotechnologies and devices would work as anticipated, and whether nanotechnologies could be controlled.	[18]
2004	1011 Japanese adults in the Tokyo area	About 55% of those respondents had heard of nanotechnology either frequently or from time to time.	Half of those surveyed believed that nanotechnology would improve their lives in the next 20 years.	88% thought positively about nanotechnology's benefit to society, but 55% were concerned that the advancement of nanotechnology could present risks to safety, unexpected outcomes, or moral issues. "[T]he level of trust in scientists in terms of the nanotechnology-related information is the highest (54%) among NGO, Industry, government, TV and other media. And the government received the lowest trust (22.5%)."	[19]

TABLE 1.1 (CONTINUED)

Public Perceptions Regarding Nanotechnology Based on Opinion Polls

Year	Population Surveyed	Knowledge of Nanotechnology	Perception of Risk	Other Findings	Ref.
2004	1536 adults randomly selected across U.S.	>80% knew little or nothing about nanotechnology.	78% thought risks and benefits were equal or benefits outweighed risks; those who knew more about nanotechnology believed benefits would outweigh risks.	Respondents did not trust business leaders to minimize nanotechnology risks to human health.	[14]
2005	177 adults in U.S. focus groups in Spokane, WA,; Dallas, TX,; Cleveland, OH	54% initially knew almost nothing about nanotechnology.	Participants opinions were surveyed. They were presented with information about nanotechnology, then surveyed again. Perceptions that benefits would be greater than or equal to risks increased from 29.4 to 75.6%; that risks would exceed benefits, from 5.1 to 15.3%. (Responses "don't know" decreased.)	After being presented with information on nanotechnology, participants held little trust in government and industry to protect the public from the risks of nanotechnology. Focus group members also felt strongly that the public needed to be better informed and, the public should have a role in decisions about investing government funds in research and in managing the risks of nanotechnology.	[16]
2006	503 people across U.S.	Not quantified in survey.	Consumers are willing to use products containing nanomaterials when the potential benefits are high, even if there are health and safety risks.	Respondents perceived that nanotechnology offered benefits on the order of the benefits from food preservatives and chemical disinfectants, albeit at lower risk.	[15]

TABLE 1.1 (CONTINUED)
Public Perceptions Regarding Nanotechnology Based on Opinion Polls

Year	Population Surveyed	Knowledge of Nanotechnology	Perception of Risk	Other Findings	Ref.
2007	1014 adults in U.S.	Approx. 70% had heard just a little or nothing at all about nanotechnology.	Roughly half of the respondents were not sure whether the benefits of nanotechnology would outweigh the risks; those who had greater knowledge of nanotechnology believed more strongly in its benefits.	In general, the group of people with little or no knowledge about nanotechnology included women, older adults, adults with a high school degree or less, and adults with lower incomes.	[17]

the potential risks of nanotechnology do not seem to be slowing the race to bring nanotechnology products to the marketplace.

1.3 ABOUT THIS BOOK

This book seeks to demystify, as much as is possible based on the current state of the science, the occupational and environmental concerns about intentionally manufactured nanomaterials. To provide context for that discussion, it begins with an explanation of nanoscale materials, their properties, and their uses, and describes the processes used to manufacture nanoscale materials. Subsequent chapters provide information on possible risks to human health and the environment and on developing regulations to manage those risks. The penultimate chapter of this book examines an apparent paradox given concerns over the possible risks of nanotechnology: the use of nanoscale materials to remediate environmental pollution.

This organizational structure — describing the manufacture of nanoscale materials, considering their usage, and following those materials through various discharges to points of exposure to gauge the consequences of exposure — parallels the process of Life Cycle Analysis (LCA). The final chapter of this book discusses frameworks such as LCA for evaluating the balance between risk and reward, and presents brief examples.

This book offers a snapshot of a rapidly developing field. It presents the current state of the science and identifies critical areas undergoing further research. Even such fundamentals as definitions, classification schemes, and understanding of the properties of nanoparticles are evolving. Intensive research into many aspects of the behavior of nanoparticles in the environment and biological systems continues, and thus the information presented in this book represents an initial framework for understanding nanotechnology and the environment. A later edition of this book, if written in a few years, would contain much more detail and could address technical questions more fully. But the community of scientists, engineers, regulators, and the public cannot wait for the results of mature research; given the possible risks and rewards of nanotechnology, we must explore what is known about the ramifications of nanotechnology and the environment now.

REFERENCES

1. United Nations Environment Programme. 2007. *GEO Year Book 2007: An Overview of our Changing Environment. Emerging Challenges — Nanotechnology and the Environment.* ISBN: 978-92-807-2768-9. http://www.unep.org/geo/yearbook/yb2007/. (Accessed February 20, 2007)
2. U.S. Environmental Protection Agency. 2007. *Nanotechnology White Paper.* EPA/100/B-07/001. Prepared for the U.S. Environmental Protection Agency by members of the Nanotechnology Workgroup, a group of EPA's Science Policy Council: 5, 22–27. http://www.epa.gov/osa/nanotech.htm. (Accessed February 21, 2007)
3. Thompson, J.A.M., J.W. McKinley, R.C. Harris, et al. 2003. MTBE occurrence in surface and ground water. In *MTBE Remediation Handbook*, Ed. E.E. Moyer and P.T. Kostecki, p. 63–70. Amherst Scientific Publishers.
4. U.S. Environmental Protection Agency. 2007. MTBE (methyl-*t*-butyl ether) in Drinking Water. http://www.epa.gov/safewater/contaminants/unregulated/mtbe.html. (Accessed June 27, 2007)
5. The Royal Academy of Engineering, the Royal Society. 2004. *Nanoscience and Nanotechnologies: Opportunities and Uncertainties*, Chapter 3. 29 July. http://www.royalsoc.ac.uk. (Accessed October 15, 2006)
6. Oakdene Hollins Ltd. 2007. Environmentally Beneficial Nanotechnologies: Barriers and Opportunities. A Report for the Department for Environment, Food, and Rural Affairs, United Kingdom (May).
7. President's Council of Advisors on Science and Technology (PCAST). 2005. The National Nanotechnology Initiative at Five Years: Assessment and Recommendations of the National Nanotechnology Advisory Panel, May 18, 2005. http://www.ostp.gov/pcast/pcast.html. (Accessed July 18, 2005)
8. U.S. Environmental Protection Agency. 2007. *Nanotechnology White Paper.* EPA/100/B-07/001. Prepared for the U.S. Environmental Protection Agency by members of the Nanotechnology Workgroup, a group of EPA's Science Policy Council: 14. http://www.epa.gov/osa/nanotech.htm. (Accessed February 21, 2007)
9. Federal Institute for Occupational Safety and Health (BAuA), Federal Institute for Risk Assessment (BfR), Federal Environment Agency (UBA), Germany. 2006. Nanotechnology: Health and Environmental Risks of Nanoparticles — Research strategy. Draft, August 2006. http://www.baua.de/en/Topics-from-A-to-Z/Hazardous-Substances/Nanotechnology/Nanotechnology.html__nnn=true. (Accessed September 26, 2007)
10. Arnall, A.H. 2003. Future Technologies, Today's Choices Nanotechnology, Artificial Intelligence and Robotics; A Technical, Political and Institutional Map of Emerging Technologies. A report for the Greenpeace Environmental Trust (July): 7. http://www.greenpeace.org.uk/MultimediaFiles/Live/FullReport/5886.pdf. (Accessed June 29, 2007)
11. Newsbank, Inc. Keyword search on "nanotechnology." http://infoweb.newsbank.com. (Accessed June 30, 2007)
12. Expanded Academic ASAP™ on InfoTrac Web by Thomson Gale. Keyword search on "nanotechnology." http://find.galegroup.com/itx/start.do?prodId=EAIM&userGroupName=mlin_c_grotpl. (Accessed June 30, 2007)

13. Woodrow Wilson International Center for Scholars. Project on Emerging Nanotechnologies: A Nanotechnology Consumer Products Inventory. http://www.nanotechproject.org/index.php?id=44&action=intro. (Accessed June 27, 2007)

14. Cobb, M.D. and J. Macoubrie. 2004. Public Perceptions about Nanotechnology: Risks, Benefits and Trust. http://www2.chass.ncsu.edu/cobb/me/past%20articles%20and%20working%20papers/Public%20Perceptions%20about%20Nanotechnology%20-%20Risks,%20Benefits%20and%20Trust.pdf (Accessed June 29, 2007.) Also published in the *Journal of Nanoparticle Research*, 6(4), 395–405.

15. Rice University Center for Biological and Environmental Nanotechnology (CBEN). 2007. Survey Shows Consumers Neutral on Risks, Benefits of Nanotechnology. Rice NSEC EEC-0647452. http://cohesion.rice.edu. (Accessed June 30, 2007)

16. Macoubrie, J. 2005. Informed Public Perceptions of Nanotechnology and Trust in the Government. Woodrow Wilson International Center for Scholars Project on Emerging Nanotechnologies. September. http://www.wilsoncenter.org. (Accessed June 30, 2007)

17. Peter D. Hart Research Associates, Inc. 2007. Awareness of and Attitudes toward Nanotechnology and Federal Regulatory Agencies. A Report of Findings Based on a National Survey among Adults. Conducted on Behalf of: Project on Emerging Nanotechnologies, The Woodrow Wilson International Center for Scholars. 25 September. http://www.wilsoncenter.org. (Accessed September 28, 2007)

18. The Royal Academy of Engineering, the Royal Society. 2004. *Nanoscience and Nanotechnologies: Opportunities and Uncertainties*, Chapter 7. 29 July. http://www.royalsoc.ac.uk. (Accessed October 15, 2006)

19. Fujita, Y. and S. Abe. 2005. Perception of Nanotechnology among General Public in Japan. Nanotechnology Research Institute, *Asia Pacific Nanotech Weekly* 4(6) http://www.nanoworld.jp/apnw/articles/library4/pdf/4-6.pdf. (Accessed June 29, 2007)

2 Nanoscale Materials
Definition and Properties

Kathleen Sellers
ARCADIS U.S., Inc.

CONTENTS

This chapter provides a working vocabulary to describe nanoscale materials. It identifies a subset of nanoscale materials that are or will potentially be in most "common" use, and describes those materials as a foundation for understanding their fate and transport and possible toxicological effects.

This chapter mentions commercially available products containing nanomaterials to illustrate various aspects of nanotechnology. Unless otherwise noted, product information in this chapter originated from the Nanotechnology Consumer Products Inventory maintained by the Woodrow Wilson Institute [1].

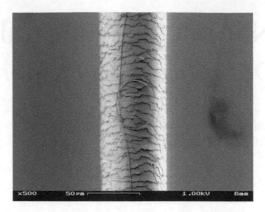

FIGURE 2.1 Micrograph of a nanowire curled into a loop in front of a strand of human hair. (From Macmillan Publishers Ltd., Tong, L. et al. 2003. Subwavelength-diameter silica wires for low-loss optical wave guiding. *Nature,* 426 (18 December): 816–819. With permission.)

2.1 DEFINITIONS

A nanometer (nm) is one billionth of a meter. This size scale can be difficult to grasp and is perhaps best understood by analogy to common materials. A nanometer is about 100,000 times smaller than either the diameter of a human hair (Figure 2.1) or the thickness of a sheet of paper. A red blood cell is approximately 5000 nm in size. Another way to grasp the relative scale of nanoparticles is this: the diameter of a fullerene, which is a spherical nanoparticle 1 nm in diameter comprising 60 carbon atoms, is approximately 10^8 times smaller than a soccer ball, which in turn is about 10^8 smaller than the planet Earth.

ASTM International [2] defines nanotechnology as "a term referring to a wide range of technologies that measure, manipulate, or incorporate materials and/or features with at least one dimension between approximately 1 and 100 nanometers (nm). Such applications exploit the properties, distinct from bulk/macroscopic systems, of nanoscale components." A nanoparticle is "a sub-classification of ultrafine particle with lengths in two or three dimensions greater than 0.001 micrometer (1 nanometer) and smaller than about 0.1 micrometer (100 nanometers) and which may or may not exhibit a size-related intensive property." The U.S. Environmental Protection Agency (U.S. EPA) cites a slightly different definition of nanotechnology: "research and technology development at the atomic, molecular, or macromolecular levels using a length scale of approximately one to one hundred nanometers in any dimension; the creation and use of structures, devices and systems that have novel properties and functions because of their small size; and the ability to control or manipulate matter on an atomic scale." [3] These definitions suggest three critical aspects of nanotechnology: (1) size, (2) functionality due to that size, and (3) intentional manufacture. As the Nanoforum notes [4], "nanotechnology should not be viewed as a single technique that only affects specific areas. It is more of a 'catch-all' term for a science which is benefiting a whole array of areas, from the environment, to healthcare, to hundreds of commercial products."

This book focuses on a subset of materials that are in or have the potential to be in the widest use at the nano scale:

- Titanium dioxide: particles of TiO_2 at the nanometer scale.
- Zero-valent iron: particles of Fe^0.
- Silver: particles of Ag.
- Carbon black: particulate form of elemental carbon.
- Carbon nanotube: hollow tube, commonly made of a single layer of carbon atoms (single-walled carbon nanotube) or multiple layers (multi-walled carbon nanotube). Nanotubes also can incorporate materials other than carbon.
- Fullerene: also known as a "buckminsterfullerene" or "buckyball," a fullerene is a hollow sphere. The term commonly refers to C60 fullerenes comprising 60 carbon atoms. Other fullerene structures, such as C70 and C120, exist.

Information about this subset of nanomaterials will provide the reader with an overview of the range of manufacturing processes, physical characteristics, and toxicological effects of nanoscale materials. Table 2.1 summarizes information about the structure and composition of the specified nanomaterials and indicates some of their uses. The focus on these materials continues through subsequent chapters of this book on the manufacture of nanomaterials, fate and transport, and potential toxicological effects.

The sections that follow provide more information about the classification, properties, and uses of common nanomaterials.

2.2 CLASSIFICATION OF NANOSCALE MATERIALS

Classification schemes used to describe nanomaterials continue to evolve, but generally recognize the origin of the material, whether it is fixed in a structure or free, its shape, and/or its composition.

2.2.1 ORIGIN

Some nanoscale materials occur naturally. Examples include volcanic ash and viruses. Human activities can generate nanoscale materials as incidental byproducts. Diesel exhaust particles and byproducts of welding, for example, can be in the nanoscale range [3, 5]. This book focuses on intentionally manufactured nanoscale materials, however, and not on these naturally occurring or incidental materials.

2.2.2 COMPOSITION AND STRUCTURE

Nanoscale materials can be made of elemental carbon, carbon-based compounds, metals or metal oxides, or ceramics. They can take many shapes. These generally include the following structures:

- Particles or crystals
- Tubes, wires, or rods

TABLE 2.1

Types and Uses of Nanoparticles

	Composition					
	Carbon		**Metals**		**Ceramic/Silica**	
Structure	**Nano-particle**	**Examples of Uses**	**Nano-particle**	**Examples of Uses**	**Nano-particle**	**Examples of Uses**
Particle	Carbon black	Pigment; reinforce-ment of rubber products	Titanium dioxide (TiO_2)	Cosmetics; environ-mental remediation	Ceramic nano-particles	Coating on photo paper
	Nano-sized wax particles	Car wax	Zero-valent iron; nano-magnetite (Fe_3O_4)	Environ-mental remediation		
			Silver	Antibacterial agent in wound care, athletic clothing, washing machines		
			Zinc oxide	Cosmetics		
			Cerium oxide	Diesel additive to decrease emissions		
Tube/wire	Carbon nanotubes	Electronics; sporting goods	Nanowire			
Dendrimer	G5 dendrimer	Targeted drug delivery	Iron sulfide clusters immobilized in dendrimers	Environ-mental remediation		
Other	Fullerene	Cosmetics	Quantum dots	Semi-conductors	Function-alized ceramic nano-porous sorbents	Water treatment

- Dendrimers (branched structures)
- Composites
- Other (e.g., spherical)

Some authorities further subdivide these general structural categories. For example, the National Institute for Occupational Safety and Health (NIOSH) defines 11 categories of nanomaterial structure [6]:

1. Agglomerated spheres
2. Colloids
3. Crystalline
4. Films
5. Nanohorns
6. Nanorods
7. Nanotubes
8. Nanowires
9. Quantum dots
10. Spherical
11. Other

Others classify nanoparticles by dimension [7–9]. According to this classification scheme, one-dimensional structures include nano films, two-dimensional nanomaterials include nanotubes, and three-dimensional shapes include fullerenes.

2.2.3 FREE VERSUS FIXED NANOPARTICLES

Free nanoparticles, as the name implies, are in solution or suspension. As a result, exposure can occur during manufacture, use, and after disposal. Commercial products containing free nanoparticles include:

- Diesel fuel containing cerium oxide to reduce emissions
- Certain sunscreens containing titanium dioxide and a face cream containing fullerenes
- Drugs containing dendrimers for targeted delivery
- Certain food products, for example vegetable oils, containing nanodrops of components such as vitamins, minerals, and phytochemicals

Free nanoparticles are likely to be released into the environment through a variety of pathways as materials are used and disposed.

Alternatively, manufacturing processes may fix nanoscale particles into a solid, as in the following examples:

- Composite tennis rackets strengthened with carbon nanotubes
- Rubber products reinforced with carbon black
- Computer chips containing nanoscale transistors

Once in use, these commercial products containing fixed nanoparticles are unlikely to release nanomaterials to the environment. Therefore, the potential for human or ecological exposure to fixed nanoparticles is limited after their incorporation into the final manufactured materials.

2.3 PROPERTIES OF NANOSCALE MATERIALS

2.3.1 OVERVIEW

The properties of nanoscale materials generally differ from those of the same materials in bulk size. This effect results from two aspects of the small size of nanoparticles: (1) the increased relative surface area per unit mass, and (2) the influence of quantum effects. Each of these points is discussed below.

2.3.1.1 Effect of Increased Surface Area

Reducing the size of a particle increases the ratio of surface area to mass. Because the reactive portion of the particle is at its surface, increasing the relative surface area will increase reactivity for a given amount of material. To illustrate, consider a spherical particle 0.1 millimeter (mm) in diameter. Its surface area is 3×10^{-8} m^2. If technicians mill the same mass of material into 100-nm-sized spheres, then the total surface area increases to 3×10^{-5} m^2. Decreasing the diameter of the particle by a factor of 1000 increases the surface area by a factor of 1000. If reactive sites cover the surface of the particle, then — all else being equal — this decrease in particle size would increase reactivity substantially. This effect accounts for the increased efficiency of nanoscale catalysts compared to their bulk counterparts.

2.3.1.2 Influence of Quantum Effects

At the nanoscale, both classical physics and quantum physics can govern the behavior of a particle. The influence of quantum effects can change essential material characteristics such as optical, magnetic, and electrical properties. An in-depth explanation of the relevant physics is beyond the scope of this book, which focuses on the implications of nanotechnology for the environment. In lieu of pages of theoretical explanation and equations, consider the following examples.

In our everyday, visible world, objects move according to Newton's models of velocity, acceleration, inertia, and momentum. For example, one can predict the trajectory of a lacrosse ball based on its mass and velocity, the pull of gravity, and the resistance of the air. If the lacrosse ball splashes into a pond, its final trajectory also will reflect the buoyancy of the water.

Other factors, however, can influence the movement of a molecule or certain nanoparticles. Even nonpolar molecules exhibit slight, transient polarity of charge because of instantaneous shifts in electron density. (Quantum mechanics projects this electron density probabilistically.) A slight negative charge on a portion of a molecule or nanoparticle due to this transient polarity will be attracted to a positive charge or repelled by a negative charge. These weak and transient intermolecular forces are called Van der Waals forces. For many nanoparticles — unlike the lacrosse ball — these slight, transient forces can influence the movement of a particle through liquid. Van der Waals forces can cause nanoparticles to agglomerate, or adsorb to each other via physisorption (physical adsorption). (For further information on the fate and transport of nanoscale materials, see Chapter 6.)

Just as Newton's laws predict the movement of a large solid, Ohm's Law relates current, voltage, and resistance to model the bulk flow of electrons through a metallic wire. Solid carbon in the form of the graphite used in pencil lead does not conduct electricity well. This property can change when a sheet of graphite one atom thick is wrapped to form a single-walled carbon nanotube. Some carbon nanotube structures can function as semiconductors. Others can conduct electricity as if the material were metal, although virtually without resistance as a result of the coordinated transfer of electrons between atoms straight down the length of the nanotube. The flow of electricity reflects individual packets of energy associated with the movement of individual electrons, rather than the bulk flow of electrons modeled by Ohm's Law.

The next example of the unique properties associated with nanoparticles is, ironically, centuries old. Medieval glass blowers used nanoscale particles of gold to color stained-glass windows. The optical properties of gold change at nanoscale. As metal particles become smaller, the quanta (or discrete packets) of light energy that can interact with them increase. Depending on their size, nanoscale gold particles can be purple, green, orange, or red [10]. Similarly, zinc oxide — notorious for coating lifeguards' noses white a generation ago — becomes transparent at the nanoscale.

These examples illustrate some of the ways in which the small size of some nanoparticles can affect their behavior and properties. It is these changes in properties relative to those of the corresponding bulk materials that account for many of the uses and much of the excitement surrounding nanotechnology.

2.3.2 CRITICAL PHYSICAL AND CHEMICAL PROPERTIES

It is clear that nanoscale materials do not necessarily behave in ways predicted from the behavior of their traditionally scaled counterparts. As a result, the physical and chemical properties that scientists usually use to predict environmental fate and transport and the consequences of exposure do not suffice to characterize nanoscale materials. Table 2.2 lists properties that may be relevant to nanotechnology and the environment according to three paradigms:

1. *U.S. EPA's voluntary Nanoscale Materials Stewardship Program (NMSP) under the Toxic Substances Control Act (TSCA) [11].* As described further in Chapter 4, the U.S. EPA has proposed this program to gather information about nanomaterials to provide a basis for developing regulations.
2. *The Voluntary Reporting Scheme for engineered nanoscale materials developed by the Department for Environment, Food and Rural Affairs (Defra) in Great Britain [13].* Defra established this program to collect information needed to assess the extent to which current regulations and controls suffice to control the potential risks from nanomaterials.
3. *The Nano-Risk Framework, which the Environmental Defense–DuPont Nano Partnership developed to evaluate the potential risks of nanoscale*

TABLE 2.2
Critical Properties of Nanomaterials

Property	Nanoscale Materials Stewardship Program [11]	Voluntary Reporting Scheme [12]	Life Cycle Analysis: NanoRisk Framework[a] [13]
1. Nomenclature:			
Technical name	•	•	•
CAS Registry Number	•	•	•
Commercial name/trade name	•	•	•
Common name	•	•	•
2. Physical/chemical properties:			
A. General characteristics:			
Chemical composition, including surface coating	•	•	•
Molecular structure	•	•[b]	•
Crystal structure	•		•
Physical state/form at room temperature and pressure	•	•	•
B. Purity of commercial product:			
Purity (or impurities in commercial product)	•	•	•
Byproducts resulting from the manufacture, processing, use, or disposal of the chemical substance	•	•	
Stabilizing agent, inhibitor, or other additives		•	
C. General properties:			
pH (at specified concentration)	•		
Solubility in water	•	•	•
Vapor pressure	•	•	•
Henry's Law coefficient	•		
Melting temperature	•	•	
Boiling/sublimation temperature	•	•	•
Flash point		•	
Self-ignition temperature		•	
Dispersability			•
Bulk density	•	•	•
Dissociation constant	•		
Surface tension		•	
Any unique or enhanced properties that arise from the nanoscale features of the material	•		

TABLE 2.2 (CONTINUED)
Critical Properties of Nanomaterials

	Paradigm		
Property	Nanoscale Materials Stewardship Program [11]	Voluntary Reporting Scheme [12]	Life Cycle Analysis: NanoRisk Framework[a] [13]
D. Particle characteristics:			
Particle size and size distribution (granulometry)	•	•	•
Aspect ratio	•		
Average aerodynamic diameter	•		
Average particle mass	•		•
Particle shape	•	•	•
Particle density	•		•
Agglomeration state	•	•	•
Deglomeration and disaggregation properties		•	
E. Surface characteristics:			
Surface area	•	•	•
Average particle surface area	•		
Surface charge/zeta potential	•		•
Porosity	•		•
Surface chemical composition	•		•
Surface reactivity			•
Surface area/volume ratio	•		
3. Production process:			
Production type (batch/continuous) and rate	•		•
Brief description of manufacturing process		•	
Source of the material, where the material is not produced by the notifier		•	
Intended uses of the material and benefits of the uses		•	
Detailed description of production process, including unit operations, chemical conversions, and mass balance (including potential releases to the environment)	•		•
Potential worker exposure	•	•	•
Personal protective equipment/engineering controls	•		•
Environmental release and disposal	•	•	•
4. Methods for characterization:			
Spectra	•	•	
Chromatographic data (high-pressure liquid chromatography, gas chromatography)		•	

TABLE 2.2 (CONTINUED)
Critical Properties of Nanomaterials

	Paradigm		
Property	**Nanoscale Materials Stewardship Program [11]**	**Voluntary Reporting Scheme [12]**	**Life Cycle Analysis: NanoRisk Framework[a] [13]**
Methods of detection and determination for the substance and its transformation products after discharge into the environment		•	
5. Environmental fate and transport[c]:			
Diffusion rate	•		
Gravitational settling rate	•		•
Sorption rate	•		•
Deposition rate	•		•
Wet and dry transport	•		•
Adsorption-desorption coefficients	•	•d	•
Octanol-water partition coefficient	•	•	
Volatilization from water	•		
Volatilization from soil	•		
Distribution among environmental media		•	
Nanomaterial aggregation or disaggregation in exposure medium of concern			•
Biodegradability (organic nanomaterials only)	•	•	•
Bioaccumulation potential	•	•	•
Biotransformation	•		
Photodegradability	•		•
Stability in water (hydrolysis)			•
Influence of redox reactions	•	•	•
Abiotic degradation		•	
6. Safety hazards:			
Flammability	•	•	•
Explosivity	•	•	•
Incompatability			•
Reactivity			•
Corrosivity			•
7. Human health hazards:			
Any hazard warning statement, label, material safety data sheet, or other information which will be provided to any person who is reasonably likely to be exposed to this substance	•		•

TABLE 2.2 (CONTINUED)
Critical Properties of Nanomaterials

	Paradigm		
Property	Nanoscale Materials Stewardship Program [11]	Voluntary Reporting Scheme [12]	Life Cycle Analysis: NanoRisk Framework[a] [13]
Acute toxicity:			
Administered orally		•	
Administered by inhalation		•	
Administered cutaneously		•	
Eye irritation		•	
Repeated dose toxicity (28 days)		•	
Short-term toxicity, including one or more of the following:			•
• 28-day inhalation study with full histopathology, over 90-day observation period			
• Single-dose instillation study with full histopathology, over a 90-day observation period			
• 28-day repeated-dose oral toxicity test with full histopathology, over a 90-day observation period			
Skin sensitization/irritation		•	•
Skin penetration, if valid tests exist			•
Genetic toxicity tests		•	•
Assessment of the toxicokinetic behavior derived from base set data and other relevant information		•	
Toxicity assessment derived from non-animal test methods, including *in vitro* methods and Quantitative Structure Activity Relationships (QSARs)		•	
8. Environmental hazards:			
Bacteriological inhibition		•	
Acute aquatic toxicity to:		•	•
• Fish (fathead minnow or rainbow trout)			
• Invertebrates (daphnia) — acute or chronic depending on conditions			
• Aquatic plants (algae)			
Terrestrial toxicity, including acute toxicity to:			•
• Invertebrates (earthworms)			
• Plants			
9. Risk management practices:			
Possibility of recycling		•	
Possibility of neutralization of unfavorable effects		•	

TABLE 2.2 (CONTINUED)
Critical Properties of Nanomaterials

	Paradigm		
Property	Nanoscale Materials Stewardship Program [11]	Voluntary Reporting Scheme [12]	Life Cycle Analysis: NanoRisk Framework[a] [13]
Possibility of destruction		•	
Others		•	
10. Existence of non-disclosed data		•	

[a] Base set data; additional data may be needed.
[b] Description and measurement of the structure of the nanoscale material, including details of the measurement technique used.
[c] In addition to physical/chemical properties.
[d] Absorption/desorption screening test.

materials [13]. Chapter 11 discusses this paradigm for Life Cycle Analysis in more detail.

Other specialized paradigms, for example, the Assay Cascade Protocol that the National Cancer Institute uses to characterize the compatibility of nanomaterials with biological systems [14], may stipulate other critical parameters. Perhaps not surprisingly, not all the data listed in Table 2.2 are readily available yet for the nanomaterials currently in industrial and commercial use.

The critical properties of nanomaterials, as listed in Table 2.2, include many "conventional" parameters that scientists use to characterize bulk chemicals. Such properties include solubility, vapor pressure, boiling point, and other phase properties; reactivity and degradability; and toxicity based on various bioassays. The critical properties also include some of particular importance to nanomaterials, as follows.

- *Particle size.* The small size of nanoparticles increases the surface area per unit volume relative to a material's bulk counterpart. The small size also affects the particles' fate and transport in the environment. Nanoparticles can generally remain suspended in air or water because their small size limits gravitational settling. As particles agglomerate and the net particle size increases, they can drop out of suspension. Particle size also affects a particle's ability to penetrate into bodily organs.
- *Particle shape.* The shape of a nanoparticle affects its ability to agglomerate and react, and to penetrate into bodily organs. "Steric hindrance" occurs when the shape of a particle or molecule physically prevents a reaction from occurring.
- *Particle surface area.* Increased relative surface area (as a result of the small particle size) increases the reactivity of nanomaterials compared to their bulk counterparts and affects other properties.

- *Explosivity, flash point, and self-ignition temperature.* The high surface area of very small particles increases their tendency to combust when suspended in air in the presence of an ignition source such as static charge or sparks. Readers may be familiar with this phenomenon from reports of dust explosions in grain elevators. The potential for combustion can be a safety issue for some nanomaterials.
- *Degree of agglomeration*.* Van der Waals forces, which are weak, transient intermolecular forces resulting from transient polarity related to shifts in electron density, can cause nanoparticles to agglomerate. Agglomeration increases the net particle size, thereby changing the size-dependent characteristics and behavior of the original nanomaterial. The Hamaker constant represents the net van der Waals attraction.
- *Surface charge.* This affects dissolution, suspension in water, and sorption, and is often represented by the zeta potential. A positive charge on the surface of a colloid (such as a metal oxide nanoparticle) in water attracts negatively charged ions in the fluid. These negatively charged ions form the so-called "Stern layer" around the colloid. The zeta potential is the charge measured at the outermost portion of the Stern layer. As discussed further in Chapter 6, the stability of a nanoparticle suspension relates to its zeta potential. The electrostatic repulsion resulting from surface charge can counter the tendency toward agglomeration.

These properties affect the fate and transport of nanomaterials in the environment, their toxicity, and their fate in wastewater treatment, as discussed in subsequent chapters.

2.4 TYPES OF NANOMATERIALS AND APPLICATIONS

New applications of nanotechnology appear constantly. The Nanotechnology Consumer Products Inventory maintained by the Woodrow Wilson Institute listed over 500 consumer products containing nanomaterials as of June 2007 [1]. The inventory grew from 212 to 502 products between March 2006 and June 2007, demonstrating the explosive growth of this market. Those products represent an extraordinarily wide range of applications. The list of products below demonstrates the range by example. (Characterizations of nanoscale ingredients in the materials on this list, and their effects, were made by the manufacturers as cited by the Nanotechnology Consumer Products Inventory. The information provided below is not intended as an endorsement, but simply to illustrate the range of products and applications.)

- Cosmetics and personal care products:
  - RevitaLift® Intense Lift Treatment Mask (L'Oreal®) — uses nanosomes, tiny capsule-like structures, to transport active ingredients into the skin's outer layer and then release them.

---

* ASTM International distinguishes between agglomeration and aggregation of nanoparticles as follows [2]. An agglomerate is a group of particles held together by relatively weak forces (such as Van der Waals forces) that can be broken apart. An aggregate is a discrete group of particles composed of individual components that are tightly bonded together and not easily broken apart.

- Serge Lutens Blusher (Barneys New York®) — "Nano Dispersion technology" creates a fine powder.
- Chemical-Free [sic] Sunscreen SPF 15 (Burts Bees®, Inc.) — contains nano-sized particles of titanium dioxide as the active ingredient.
- Food supplements and food storage:
  - MesoZinc™ (Purest Colloids, Inc.) — nutritional supplement containing 30 parts per million (ppm) zinc nanoparticles.
  - FresherLonger™ Miracle Food Storage (Sharper Image®) — food storage containers are infused with silver nanoparticles as an antibacterial agent.
  - Silver Nano Baby Milk Bottle (Baby Dream® Co., Ltd.) — "silver nano poly system" acts as an antibacterial and deodorizer.
- Appliances:
  - Samsung® Washing Machine (Samsung®) — Silver Nano technology "sterilizes your clothes."
  - Daewoo® Vacuum Cleaner (Daewoo®) — nano-silver coated cyclone canister removes bacteria.
  - Samsung® Air Conditioner — contains silver nano filter and silver nano evaporator.
- Clothing:
  - Sport Anklet Sock (AgActive) — treated with nanoparticles of silver (typically 25 nm) as bactericide and fungicide.
  - NANO-PEL™ clothing (Nordstrom® Inc.) — fabric used in clothing such as pants is treated with Nano-Tex process to bind water-repellent molecules to cotton fibers, in order to impart stain resistance.
- Coatings:
  - Pilkington Activ™ Self Cleaning Glass (Pilkington plc) — glass coating that works with ultraviolet (UV) light and rain to keep glass free from organic dirt.
  - Turtle Wax® F21™ Car Wash (Turtle Wax®, Inc.) — nanotechnology formula comprising synthetic polymers provides protection against UV light.
  - Behr® PREMIUM PLUS® Exterior Paint (Behr® Process Corporation) — proprietary nanoparticles improve adhesion and anti-mildew properties.
  - Ultima® Photo Paper (Eastman Kodak® Company) — nine-layer composition incorporates ceramic nanoparticles to resist the effects of heat, humidity, light, and ozone.
- Electronics and computers:
  - Invisicon™ (Eikos®) — Invisicon™ ink used to create transparent conductive coatings and manufacture printed circuits on transparent plastic films; Invisicon™ incorporates carbon nanotubes with a 1000:1 aspect ratio. Applications include flat panel displays.
  - XBOX 360® (Microsoft®) — microprocessor chip manufactured by IBM using IBM's 90 nanometer Silicon on Insulator (SOI) technology to reduce heat and improve performance.

- Sporting goods:
  - Wilson® Tour Davis Cup Official Tennis Ball (Wilson®) — incorporates "NanoPlay" technology to increase durability.
  - Head® Nano Titanium Tennis Racquet (Head®) — integrates nanoscale materials.
  - Wilson® [K]Factor® Tennis Racket (Wilson®) — contains Karophite Black, created by bonding carbon black, graphite, and silicon dioxide together at the nano level.

Some of these product descriptions illustrate how manufacturers can guard the details of proprietary technology by providing little information about the nanomaterials in their products. Many of these consumer products do, however, contain one or more of the nanomaterials that are the focus of this book. The sections below provide information on titanium dioxide, zero-valent iron, silver, carbon black, carbon nanotubes, and fullerenes.

2.4.1 TITANIUM DIOXIDE

Titanium is a common element, found in the minerals rutile (predominantly titanium dioxide, TiO_2, also known as titania) and ilmenite ($FeTiO_3$). Manufacturers have long used TiO_2 as a pigment, and in welding electrodes, ceramics, and catalysts. As the primary white pigment, it is used to color paint, plastics, paper, and inks. The functions of TiO_2 depend on its ability to absorb or reflect light of different wavelengths. TiO_2 acts as a pigment because the particles can scatter visible light, a function that depends on particle size. TiO_2 also can absorb UV radiation. As a result, coatings containing TiO_2 can provide protection from photochemical degradation. Upon absorption of UV light, TiO_2 can generate hydroxyl radicals. The TiO_2 pigment then acts as a photocatalyst [15] for the decomposition of organic compounds. The crystalline structure of nanoscale TiO_2 particles allows those particles to absorb visible light as well as UV light, which broadens the applications for TiO_2 catalysts [16].

Commercially available products that exploit these properties of nanoscale TiO_2 include [1]:

- Sunscreens by more than a dozen manufacturers rely on the ability of TiO_2 to absorb UV radiation to protect the skin.
- T-2® Photocatalyst Environment Cleaner (T-2®) — when light strikes nanoscale TiO_2 in this cleaner after it is applied to a surface, photocatalytic reactions degrade "organic toxins, odors, and more."
- Carrier® Pure Dew Filtration (Carrier®) — this air purifier contains a Nano Silver filter, for antibacterial action, and a Nano Photocatalytic filter. The latter contains nano-sized particles of TiO_2 to "get rid of unpleasant smell and smoke."

2.4.2 ZERO-VALENT IRON

Engineers use nanoparticles of elemental iron known as nano zero-valent iron (nZVI), as discussed further in Chapter 10, to treat groundwater containing chlorinated solvents, arsenic, and other contaminants *in situ*. Granular ZVI has been used since the mid-1990s in groundwater remediation. Practitioners realized early in the use of granular ZVI that the rate of reductive dechlorination of chlorinated hydrocarbons depended strongly on the surface area of the ZVI particles. This realization led to the development of nZVI. Particles of nZVI generally range between 40 and 300 nm in diameter. This particle size, because of the increased surface area, provides greater reactivity than the granular ZVI initially in use. The direct injection of nZVI, however, can be limited in the field by the tendency of the nanoparticles to agglomerate, and as discussed in Chapter 10, alternative methods to alleviate this tendency are being developed.

2.4.3 SILVER

Some 95% of silver is used in photography, jewelry and silverware, and in various industrial applications. The latter includes electrical components, brazing alloys and solders, bearings, catalysts, miniature batteries, photovoltaic cells, and other products [17, 18].

Silver has long been recognized as an antibacterial agent. Cyrus the Great, King of Persia, reportedly kept water fresh in the sixth century B.C. by boiling the water and then storing it in silver flagons [19]. Pliny the Elder, writing in 78 A.D., said that silver slag "… has healing properties as an ingredient in plasters, being extremely effective in causing wounds to close up." In modern time, silver has been used to purify drinking water, sanitize swimming pools, and prevent sepsis in wounds [17, 18]. Silver's antibacterial function results from its ability to disturb the multiplication function of bacteria [17].

Thousands of years after the first apparent use of silver as a bactericide, nanoscale particles of silver lend antibacterial qualities to new commercial products. Approximately 20% of the available nanotechnology consumer products are those containing nano-scale silver. Most incorporate silver due to its antibacterial properties. These products include, as described above, clothing, food storage containers, and appliances. They also include wound care products [1].

Nano-scale silver may find more uses in the future. In one type of application, nano-scale silver is used in an extraordinary analytical technique. Two types of nanoprobes have been developed under the sponsorship of the U.S. Department of Energy [20, 21]. Each probe consists of a silica optical fiber between 20 and 100 nm in diameter coated with a thin layer of silver. The "nanobiosensor" probe allows scientists to physically probe inside a living cell without destroying it. In one application, a bioreceptor molecule, such as an antibody, DNA, or enzyme that will bind to a specific target molecule of interest inside the cell, is immobilized on the silver surface at the nanoprobe tip. Only the target molecules that become bound to the bioreceptor are exposed to and excited by an evanescent laser signal, giving off detectable fluorescent light. The team leader, Tuan Vo-Dinh, has stated that:

> "[T}he nanobiosensor has important implications ranging from drug therapy to national security, environmental protection, and a better understanding of molecular biology at a systems level. This area of research is truly at the nexus of nanotechnology, biology, and information technology."

A variation on the nanoprobe is based on the light-scattering technique known as surface-enhanced Raman spectroscopy (SERS). This probe can detect and analyze chemicals, including explosives and drugs on surfaces, at a theoretical single-molecule level. This capability makes the nanoprobe far more selective, sensitive, and accurate than conventional analytical techniques. The SERS nanoprobe is being developed for nanotechnology applications from military and water monitoring applications to medical environments.

2.4.4 CARBON BLACK

Carbon black comprises fine particles of elemental carbon. It is arguably one of the oldest nanomaterials in commercial use and certainly the most widely used. The Chinese manufactured an impure form of carbon black more than 3500 years ago by burning vegetable oils in lamps. Modern manufacturers still produce carbon black from the incomplete combustion or thermal decomposition of hydrocarbons (usually heavy aromatic oils). Worldwide, manufacturers produced approximately 18 billion pounds of carbon black in 2004. About 90% of this output found use in rubber products, where it serves as a filler and a strengthening or reinforcing agent. Carbon black also is used in plastics, coatings, and inks [22, 23].

Carbon black consists of more than 97% elemental carbon. Carbon black particles, or nodules, comprise stacks of graphite-like sheets of carbon. Carbon black products are characterized by their structure, size, surface area, surface activity, and particle size [22].

Carbon black particles generally range in size from tens to a few hundred nanometers. As a result, many classify carbon black as a nanomaterial (e.g., [3, 22, 24]). The International Carbon Black Association, however, holds a different view [23]; it holds that while the initial particle size may be on the nanoscale, van der Waals forces cause these particles to rapidly aggregate into "basic indivisible entities" of carbon black approximately 85 to 500 nm in size. These aggregates then adhere to form stable agglomerates on the order of 1 to over 100 μm in size. Manufacturers shape these aggregates into carbon black pellets between 0.1 and 1.0 mm in size before shipment to their customers. The International Carbon Black Association indicates that nanoparticles of carbon black "are not found outside the reactor, nor are they found as a component dust fraction in final manufactured carbon blacks."

These differences in the categorization of carbon black illustrate two important points regarding nanotechnology:

1. Strong attractive forces can cause individual particles to agglomerate readily, changing their characteristics and limiting their transport in the environment. (See Chapter 6 for further discussion of the forces that govern the behavior of nanoparticles.)
2. Literature reports on nanotechnology warrant careful reading to ascertain the actual particle size and form under discussion, and to determine whether experimental conditions correspond accurately to the form of a material that is actually commercially available or used. (Chapter 8 discusses some of the techniques used to suspend nanoparticles in solution for

toxicity testing, and how those techniques should affect the interpretation of the results.)

2.4.5 CARBON NANOTUBES

The properties of nanotubes sound like science fiction; these tiny cylinders can have extraordinary strength and unusual electrical properties. Their characteristics depend on their composition, size, and orientation, as described below [25–27].

This discussion focuses on carbon nanotubes (CNTs) rather than metal-based nanotubes. CNTs consist of one or more thin sheets of graphite one atom thick, known as graphene, which are rolled to create a hollow cylinder (Figure 2.2). A single-walled carbon nanotube (SWNT) contains one layer of graphene; a multi-walled carbon nanotube (MWCNT) comprises concentric cylinders of graphene. SWNT diameters generally range from 0.4 to 2.5 nm; MWCNT, up to several hundred nanometers.

In graphene, each carbon atom bonds to three other carbon atoms. The resulting hexagonal lattice resembles a honeycomb. Depending on the orientation of the hexagons — that is, on the chirality of the graphene sheet — an SWNT can take on different configurations. These different configurations affect the electrical properties of a nanotube.

In general, CNTs have the following properties:

- Insoluble in water and, for most forms of carbon nanotubes, insoluble in solvents.
- Great physical strength and flexibility. Carbon nanotubes have a Young's modulus of approximately 1 terapascal (TPa) and tensile strength up to 150 GPa. By these measures, carbon nanotubes are approximately 200 times as strong as steel.
- Light weight (density 2.6 g/cm^3, about one-third the density of steel).
- Depending on the structure, CNTs are conductors or semiconductors. A nanotube with the carbon atoms arranged in straight lines along the length of the tube conducts electricity very efficiently. When the carbon atoms are arranged in a spiral, the nanotube is a semiconductor.
- High thermal conductivity.
- High storage capacity for chemical substances within the hollow cylinder.

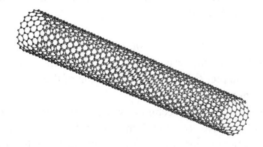

FIGURE 2.2 Illustration of a carbon nanotube. (From Chris Ewels, www.ewels.info. With permission.)

Scientists can alter the properties of CNTs by adding functional groups to the outside of the nanotubes. For example, adding hydroxyl groups to the outer surface can make CNTs water soluble.

The physical properties of CNTs may lead to a wide variety of commercial applications. CNTs may ultimately have the following uses [27]:

- Materials and chemistry applications: polymer CNT composites, coatings, membranes, and catalysis
- Medicine and life science tools: "lab on a chip" for medical diagnoses, drug delivery, chemical sensors, and filters for water and food treatment
- Electronics and information technology: lighting elements, single-electron transistors, molecular computing and data storage, electromechanical sensors, micro-electro-mechanical systems (MEMS)
- Energy: super capacitors for energy storage, solar cells, fuel cells, and superconductive material

Products containing CNTs that are now available [1] include a variety of sporting goods that utilize nanotubes to strengthen the resins. These include baseball bats, bicycle frames, golf clubs, hockey sticks, and tennis rackets. CNTs also have been used to strengthen plastics used in automobiles. Finally, two products use carbon nanotubes in video displays.

2.4.6 FULLERENES

A fullerene is a spherical particle most commonly comprising 60 carbon atoms arranged as 20 hexagons or 12 pentagons (Figure 2.3). The diameter of a C60 fullerene is approximately 10.0 Angstroms (Å), or 1.00 nm. The structure of a C60 molecule resembles one of the geodetic domes built by architect R. Buckminster Fuller, and thus the name "buckminsterfullerene," abbreviated "fullerene." The discovery

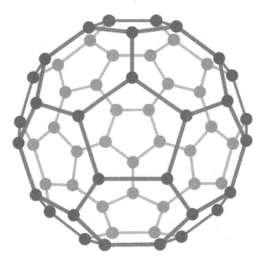

FIGURE 2.3   Fullerene.

of fullerenes merited a Nobel Prize award to Dr. Richard Smalley, Dr. Robert Curl, and Dr. Harold Kroto in 1996 [28]. Much of this work occurred at Rice University in Texas, leading the legislature of the State of Texas to designate, in 1997, that the buckminsterfullerene is the official State Molecule of Texas [29].

The C60 fullerene is the most common form, followed by C70; higher fullerenes include C74, C76, C78, C80, C82, C84, and C86 through C96. Variations on the fullerene structure include [30]:

- *Endohedral or incarfullerenes.* These fullerenes entrap (or *incar*cerate) elements or molecules within the carbon matrix.
- *Multi-walled or nested fullerenes.* These particles comprise two or more nested fullerenes.
- *Heterofullerenes.* In these fullerenes, an atom such as boron, nitrogen, or phosphorous replaces one or more carbon atoms in the fullerene.

The properties of fullerenes depend on the structure, derivatization, and degree of agglomeration. In general, pure fullerenes have low solubility in water. Under certain conditions, fullerenes will form polymorphic hexagonal unit cell agglomerates in water referred to as nano-C60. These agglomerates, approximately 25 to 500 nm in size, carry a strong negative charge. Scientists have detected fullerenes in ancient geologic formations, suggesting that they can originate from natural sources as well as manufacturing processes, and that fullerenes can be extraordinarily stable under geologic conditions [31].

Fullerenes are electron deficient and can react with nucleophilc species. They are not oxidized readily, although in the presence of oxygen and light, oxidative degradation of C60 to $C_{120}O$ occurs. The larger fullerenes are less reactive than C60 [30].

Possible applications for fullerenes include catalysts and sensors; nanocomposites containing fullerenes can be used in optics and photochemistry applications, and endohedral fullerenes also can be used as medical therapeutic agents [32]. Of the eight commercial products listed in the Nanotechnology Consumer Products Inventory [1] as containing carbon fullerenes, seven are cosmetics products and one is sports equipment. Some of the cosmetic products tout the ability of C60 fullerenes to scavenge free radicals.

2.5 SUMMARY

Nanomaterials share the characteristic that they are between 1 and 100 nm in at least one dimension. As a result of this small size, the physical and chemical properties of nanoparticles can differ from those of the same material in bulk form. The smaller size means that the relative surface area of the particle is larger and therefore its relative reactivity increases. For some materials, the ability of an atom or molecule to absorb or emit energy in quanta begins to influence the nanomaterial's behavior.

Beyond those generalizations, however, nanomaterials vary widely in terms of their composition, properties, and uses. To bring focus to discussions of nanotechnology and the environment, this book focuses on six engineered nanomaterials:

1. Titanium dioxide
2. Silver
3. Zero-valent iron
4. Carbon black
5. Carbon nanotubes
6. Fullerenes

REFERENCES

1. Woodrow Wilson International Center for Scholars. Project on Emerging Nanotechnologies: A Nanotechnology Consumer Products Inventory. http://www.nanotechproject.org/inventories/consumer/. (Accessed June 27, 2007)
2. ASTM International. 2006. Designation: E 2456 – 06. Standard Terminology Relating to Nanotechnology.
3. U.S. Environmental Protection Agency. 2007. *Nanotechnology White Paper.* EPA/100/B-07/001. Prepared for the U.S. Environmental Protection Agency by members of the Nanotechnology Workgroup, a group of EPA's Science Policy Council, 5, 7, 55–60.
4. nanoforum.org, undated. European Nanotechnology Gateway. What is Nanotechnology? http://nanoforum.org/#. (Accessed December 9, 2007)
5. Gwinn, M.R. and V. Vallyathan. 2006. Nanoparticles: health effects — pros and cons. *Environ. Health Perspect.,* 114(12):1818–1825.
6. National Institute for Occupational Safety and Health. Nanoparticle Information Library. http://www2a.cdc.gov/niosh-nil/index.asp. (Accessed June 27, 2007)
7. Sweet, L. and B. Strohm. 2006. Nanotechnolgy — Life-cycle risk management. *Hum. Ecolog. Risk Assess.,* 12:528–551.
8. Environmental Defense and DuPont. 2007. Nano Risk Framework. 21 June. http://www.NanoRiskFramework.com. (Accessed June 27, 2007), 13.
9. The Royal Academy of Engineering, the Royal Society. 2004. Nanoscience and Nanotechnologies: Opportunities and Uncertainties, Chapter 3. 29 July. http://www.royalsoc.ac.uk. (Accessed October 15, 2006)
10. Ratner, M. and D. Ratner. 2003. *Nanotechnology: A Gentle Introduction to the Next Big Idea,* 12–13, 35. Upper Saddle River, NJ: Prentice Hall Professional Technical Reference.
11. Environmental Protection Agency. 2007. Nanoscale Program Approach for Comment: Information Collection in Support of EPA's Stewardship Program for Nanoscale Materials; Reporting Form. Attachment A: Nanoscale Materials Stewardship Program Data Form. 12 July. http://epa.gov/oppt/nano/nmspfr.htm. (Accessed July 13, 2007)
12. Department for Environment Food and Rural Affairs, 2006. Voluntary Reporting Scheme for Engineered Nanoscale Materials — Data Reporting Form. October 2006. http://www.defra.gov.uk/ENVIRONMENT/nanotech/policy/index.htm. (Accessed September 7, 2007)
13. Environmental Defense – DuPont Nano Partnership. 2007. Nano Risk Framework. 21 June. http://www.nanoriskframework.com. (Accessed June 27, 2007)
14. U.S. National Institutes of Health, National Cancer Institute, Nanotechnology Characterization Laboratory. Assay Cascade Protocols. http://ncl.cancer.gov/working_assaycascade.asp.
15. Fisher, J. and T.A. Egerton. 2001. Titanium Compounds, Inorganic. In *Kirk-Othmer Encyclopedia of Chemical Technology.* New York: John Wiley & Sons, Inc.
16. Li, H., S.G. Sunol, and A.K. Sunol. 2005. Environmentally Friendly Pathways for Synthesis of Titanium Dioxide Nanoparticles. In *Technical Proceedings of the 2005 NSTI Nanotechnology Conference and Trade Show,* 2:62–26.

17. Etris, A.F. 2001. Silver and silver alloys. In *Kirk-Othmer Encyclopedia of Chemical Technology,* 4:761–803. New York: John Wiley & Sons, Inc.
18. The Silver Institute. 2007. Uses. http://www.silverinstitute.org/uses.php. (Accessed September 9, 2007).
19. Baker, M.N. 1948. *The Quest for Pure Water,* 4. New York: The American Water Works Association, Inc.
20. Vo-Dinh, T., P. Kasili, and M. Wabuyele. 2006. Nanoprobes and nanobiosensors for monitoring and imaging individual living cells. *Nanomed.: Nanotechnol. Biol. Med.,* 2:22–30.
21. Oak Ridge National Laboratory. 2004. News Release: ORNL nanoprobe creates a world of new possibilities. 14 July.
22. Wang, M.-J., C.A. Gray, S.A. Resnek, K. Mahmud, and Y. Kutsovsky. 2003. Carbon black. In *Kirk-Othmer Encyclopedia of Chemical Technology,* 4:761–803. New York: John Wiley & Sons, Inc.
23. International Carbon Black Association. 2004. *Carbon Black User's Guide.* http://carbon-black.org. (Accessed August 30, 2007)
24. Donaldson, K. et al. 2005. Combustion-derived nanoparticles: a review of their toxicology following inhalation exposure. *Part. Fibre Toxicol.,* 2(21 October):10. http://www.particleandfibretoxicology.com/content/2/1/10.
25. Dresselhaus, M., G. Dresselhaus, P. Eklund, and R. Saito. 1998. Carbon nanotubes. *Physics World* (January). http://physicsweb.org/articles/world/11/1/9/1. (Accessed June 22, 2007)
26. Helland, A., P. Wick, A. Koehler, K. Schmid, and C. Som. 2007. Reviewing the Environmental and Human Health Knowledge Base of Carbon Nanotubes. *Environmental Health Perspectives* (National Institutes of Health). DOI: 10.1289/ehp.9652. http://dx.doi.org. (Accessed May 10, 2007)
27. Helland, A., P. Wick, A. Koehler, K. Schmid, and C. Som. 2007. Supplemental material to manuscript: reviewing the Environmental and Human Health Knowledge Base of Carbon Nanotubes. *Environmental Health Perspectives* (National Institutes of Health). http://dx.doi.org. (Accessed May 10, 2007)
28. Royal Swedish Academy of Sciences. 1996. Press Release: The Nobel Prize in Chemistry 1996. 9 October. http://nobelprize.org/cgi-bin/print?from=/nobel+prizes/chemistry/laureates/1996/press.htm. (Accessed September 9, 2007)
29. Hochberg, S. 1997. House Concurrent Resolution H.C.R. No. 83, 75R6204 JTR-D. http://www.legis.state.tx.us/tlodocs/75R/billtext/html/HC00083.htm. (Accessed September 9, 2007)
30. Taylor, R. 2002. Fullerenes. In *Kirk-Othmer Encyclopedia of Chemical Technology,* 12:228–258. New York: John Wiley & Sons, Inc.
31. Nowack, B. and T.D. Bucheli. 2007. Occurrence, behavior and effects of nanoparticles in the environment. *Environ. Pollut.,* 150:5–22.
32. Lyon, D.Y., L.K. Adams, J.C. Falkner, and P.J.J. Alvarez. 2006. Antibacterial activity of fullerene water suspensions: effects of preparation method and particle size. *Environ. Sci. Technol.,* 40(14):4360–4366.

3 Overview of Manufacturing Processes

Julie Chen
University of Massachusetts, Lowell

Kathleen Sellers
ARCADIS U.S., Inc.

CONTENTS

This chapter describes the processes used to manufacture nanomaterials and the anticipated evolution of those processes. This information provides a basis for understanding the potential for worker exposure and environmental releases. The discussion begins with context on manufacturing processes and how they can convey desired properties to a product.

3.1 INTRODUCTION

3.1.1 Manufacturing: Form and Function

The ultimate objective of manufacturing is to impart the desired *form and function* into a product. For example, photolithography is one of several steps used to impart physical connections and electronic properties into the integrated circuit chips prevalent in everything from cell phones and computers to the latest automatic coffee

FIGURE 3.1   Example of a micro-injection molded medical implant, next to a penny for scale. (From Miniature Tool and Die, Charlton, MA, www.miniaturetool.com. With permission.)

makers. The manufacturing process must control both the geometry, in terms of the size, shape, and interconnection of components, and the presence of conducting and insulating materials in specific locations. Injection molding, a very different process from lithography, is used to make everything from large appliance and electronics enclosures to medical implants (Figure 3.1). For the latter, form is represented by the control of the implant geometry, and function by the necessary strength, stiffness, and wear properties of the material.

3.1.2   LOOKING FORWARD...LOOKING BACK

Over the many centuries of human development, the fabrication of products has changed enormously, in terms of materials, tools, scale, complexity, and degree of human interaction. However, these changes have not been purely monotonic in their progression. For example, early materials were all "natural materials" — that is, wood from trees, skins from animals, stones from the ground. Although mixing of materials to form metal alloys was conducted more than 4000 years ago, remarkable advances have been made in materials processing within the most recent 50 years. Included among these advances have been discoveries leading to new "man-made" or synthetic developments, such as shape memory alloys that change shape at a specified temperature, used for applications as varied as orthodontic wires, medical insertion devices, and military actuators; polymer fibers for ballistic protection or moisture-wicking athletic clothing; and semiconductor materials that form the core of all current electronic devices. More recently, however, there has been a return to "natural materials" in efforts to create environmentally benign materials derived from biodegradable and renewable resources.

In a similar manner, the level of skill and interaction of the worker with the product has undergone cyclic changes. Prior to the Industrial Revolution, manufacturing essentially consisted of individual hand work performed by skilled laborers. The development of mass production in the early 1900s led to a rise in unskilled labor, as manufacturing equipment developments and scientific management taken to an extreme reduced the worker to simply another component or "cog" in the assembly line. The subdivision of labor to simple motions repeated over and over again was promoted by Frederick Winslow Taylor [1]. Variations on the scientific management theme with a greater emphasis on worker welfare and mass production of products affordable by the general public were studied by Frank and Lillian Gilbreth [1] and Henry Ford, respectively. Worker conditions and the hazards of extreme industrial efficiency was a theme of Charlie Chaplin's movie *Modern Times* (1936). With new advances in automated equipment and computer control, however, the degree of repetitive assembly and inspection has decreased, and there has been a shift to skilled (albeit not in hand work) workers familiar with computers and an increase in the need for more technically knowledgeable workers.

A growing concern in more recent times is the exposure of workers to potentially hazardous environments — ranging from the obvious hazards of large mechanical and electrical equipment (e.g., crushing, falls, electrocution), to the less visible dangers of exposure to chemicals and airborne particles (e.g., coal dust). Improved safety protocols, safety lock-out systems and guards, and personal protective equipment (e.g., gloves, masks, ventilation) have been developed to address work environment hazards. Nevertheless, with the emergence of each new technology comes the potential for new, unknown hazards. Some hazards arise from the materials themselves, as in the case of asbestos fibers and lead. Others arise from the manufacturing process, as in the increase of carpal tunnel and other repetitive motion injuries. To mitigate the potential harm, the scientific community must attempt to address potential hazards prior to or in parallel with new technology development. One approach to doing so for manufacturing processes is to first identify what changes are anticipated in the manufacturing environment due to the emerging technology, and then address any subsequent consequences. As was illustrated previously, however, projecting forward is not simply a linear extension of observations of the past. For example, it is unlikely that preventing inhalation of nanoparticles will be solved solely by creating masks with smaller pores. Thus, the next section provides a brief introduction to existing manufacturing processes, followed in the ensuing section by a discussion of how these processes are likely to change with the increased use of nanomaterials.

3.2   A BRIEF PRIMER ON MANUFACTURING PROCESSES

While there are many different major processes, each with many variations, manufacturing processes can be loosely grouped into the following five families [2]:

FIGURE 3.2 Examples of mass change — material removal manufacturing processes: (a) laser machining and (b) waterjet cutting. ([a] From the Center for Lasers and Plasmas for Advanced Manufacturing (CLPAM) website, www.engin.umich.edu/research/lamircuc; and [b] from Flow International Corporation, Kent, WA, www.flowcorp.com. With permission.)

FIGURE 3.3 Example of a complex three-dimensional geometry fabricated using a mass change process — inkjet printing, an additive manufacturing process. (From Digital Design Fabrication Group, MIT Department of Architecture, http://ddf.mit.edu. With permission.)

1. *Mass change processes.* These processes involve the addition or subtraction of material. The most obvious of these is machining, which includes many methods. In addition to standard mechanically based machine tools such as drills, lathes, milling machines, and saws, other types of energy have been harnessed for material removal, including laser machining, water jet cutting, and electrodischarge machining (EDM) (Figure 3.2). Additive processes range from methods as old as electroplating, which involves using an electric current to deposit a metal coating onto a conductive substrate, to newer approaches expanding on rapid prototyping methods such as ink-jet or three-dimensional printing, selective laser sintering, and stereolithography. The rapid prototyping processes can build up complex three-dimensional shapes on a layer-by-layer basis (Figure 3.3), using advanced computer control to precisely place powders and fuse or sinter them, or to selectively cure polymers in specified locations.

2. *Phase change processes.* These processes involve the shift of the material from one phase to another (e.g., liquid to solid, vapor to solid). The initial phase provides ease of handling. For example, in injection molding, molten polymer is able to flow into small channels and features, and then solidify into a rigid part. Similarly, in casting, molten metal can be forced to fill complex geometries. Less familiar perhaps are the vapor-to-solid processes such as chemical vapor deposition (CVD) or physical vapor deposition (PVD). In these processes, energy is used to transform the desired material

into a vapor or plasma form, which is then deposited onto the substrate, typically in a thin film.

3. *(Micro-)structure change processes.* Most often used to modify properties rather than geometry, structure change processes typically involve heat treatment to remove residual stresses, increase ductility, and/or harden surfaces (e.g., precipitation hardening). The process can be used as an intermediate step in combination with other processes such as forging to enhance the ability to create the desired geometry without fracturing the material. A more recent variation on these processes is ion implantation, which is used extensively in the semiconductor industry. The implantation of small amounts of impurity atoms changes the chemical structure and thus the electronic and physical properties of the material.

4. *Deformation processes.* These processes require some level of ductility in the material. Constant cross-sections such as sheet, rod, tube, etc. can be extruded through a die of the desired shape. Other geometries can be created by matched die molding, forging, thermostamping, etc. In addition to creating the desired shape, the process can be used to modify the material, typically hardening the material with repeated impacts, such as in forging. For metals, many deformation processes are combined with structure change processes. The material is softened with heat (annealing) to increase its ductility both before deformation and after to reduce residual stresses.

5. *Consolidation processes.* Typically used for materials that are brittle and have high melting temperatures, consolidation processes are commonly used for ceramics and high melt temperature metals. The materials are initially in a powder form, which is then combined with a liquid to produce a slurry that flows into the mold. Pressure and heat are then used to compact the material and sinter the powders together to obtain strength.

The choice of manufacturing process, or in some cases the creation of new processes, depends on a multitude of factors, including geometry, dimensional tolerance, number of parts, and material. Examples of some common design decision-making aspects are:

• *Geometry: complex vs. simple.* Shapes requiring constant cross-sections can be made in continuous production, usually by forcing material through a die of the desired cross-section. For example, electrical wires are coated with insulation by forcing the conductive copper wires through a slightly larger circular hole in the presence of a molten polymer, which forms a thin coating on the wire. Similarly, large aluminum I-beams, channels, pipes, and rods are extruded in continuous production. Pulling instead of pushing is necessary for fiber-reinforced composites; hence the variation is pultrusion. One step up in complexity is the fabrication of simple but not constant cross-section geometries. These shapes can be formed easily using an automated version of the blacksmith's craft of pounding horseshoes out of rods of heated steel. At some point, however, forging, stamping, and other mechanical deformation methods become too unwieldy a technique to obtain highly complex, intricate shapes. Thus, processes that rely on fluid flow, such as casting and injection molding, are used to fabricate the many intricate parts in a model car kit or in a medical device. Other techniques that rely on a "writing"-type, layer-by-layer process also provide increased control for three-dimensional structures.

- *Dimensional tolerance and surface finish.* The importance of the dimensional precision and the surface finish affects the type of manufacturing process selected. For example, vacuum forming, which uses a rigid tool on one side and a flexible surface on the other, is a much cheaper, lower force, and more forgiving process than forming with a pair of matched die molds, but the produced part can have much greater thickness variations and surface roughness. Products such as automotive body panels require a "Class A" surface finish that displays no scratches, dimples, wrinkles, or other defects that would detract from the high-luster, polished appearance. These panels, however, only require such a finish on one side of the part — for example, no one looks at the underside of the hood. Cast parts typically have poor surface finish and dimensional tolerance because of the shrinkage and porosity that occurs as the molten metal cools. Polymers also tend to shrink significantly upon cooling; thus, many parts requiring strict dimensional control utilize filled polymers — that is, polymers mixed with short chopped fibers or other fillers — to reduce shrinkage, moisture absorption, and creep.
- *Number of parts.* The anticipated volume of parts and desire for flexibility in design play an important role in process selection. Expensive tooling, costing on the order of tens of thousands of dollars and up, is only practical if the cost can be spread over many parts. In contrast, customizable products must rely on easily modified processes such as machining and rapid prototyping. Another example can be found within the many variations on the casting process — sand casting, lost wax or investment casting, die casting, centrifugal casting, etc. The first two variations involve destroying the mold for each part, whereas the latter two variations utilize reusable molds. Reusable molds are fabricated from much more expensive materials and only become economical for the production of a large number of parts (or fewer but more expensive parts).
- *Material.* In the field of materials engineering, a common description of the interrelation of multiple factors is the structure-property-processing triangle (Figure 3.4). The design flow does not have a single starting point, as each node affects the other two. For example, the rate at which a polymer is extruded and cools affects its crystallinity (structure), which then affects its stiffness and strength (property). A material that is brittle (property) would not be suitable for forging (processing).

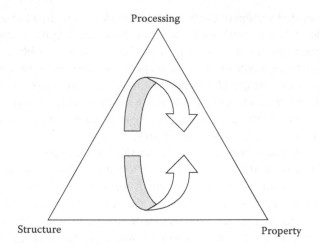

FIGURE 3.4   Structure-Property-Processing interrelationship for materials.

3.3   RAMIFICATIONS OF WORKER EXPOSURE AND ENVIRONMENTAL ISSUES FOR NANOMANUFACTURING

In considering the progression of manufacturing processes with respect to the work environment, there has been a general trend over the past hundred years toward improved safety, with significant advances made in the major industries. The rate of change, however, can vary by industry. Industries with a long history and large, expensive capital equipment naturally tend to move more slowly than newer industries that germinated with computerized, automated equipment. For example, much of the forging, casting, and sheet metal industry is still represented by workplaces that are loud, hot, and particulate-laden. In contrast, the biotechnology industry relies on clean, well-controlled environments, where the risk is more of the unseen, in both process and waste streams. Because nanotechnology and nanomaterials are anticipated to affect both of these industries and many more, the question arises as to how the manufacturing environment will change. How will issues of worker exposure and environmental impact differ for nanomaterials?

3.3.1   FOUR "GENERATIONS" OF NANO-PRODUCT DEVELOPMENT

In the case of the incorporation of nanomaterials into products, several generations of changes to manufacturing can be anticipated. Current products in the marketplace today typically fall into the "1st generation," where relatively minor modifications to existing processing equipment were needed to incorporate nanomaterials into the product. For example, surface coatings of nanofibers and nanowhiskers have been used for improved filtration and for the "nano-pants" fabric made by Nano-Tex [3]. More than 20 years ago, Toyota incorporated clay nanoparticles into polymer resins to create automotive body panels with improved strength, toughness, and dimensional stability [4]. These types of nanocomposite products are still fabricated using conventional injection molding, extrusion, and cast film processes, but additional compounding steps or other modifications to the processes were made to create a well-dispersed nanofiller [5]. As greater understanding is achieved, more advanced processes and products are developed.

The following generational designations have been described on several occasions by M.C. Roco, who is recognized as one of the key architects of the National Nanotechnology Initiative (NNI). A more detailed presentation can be found in a chapter by Roco reviewing the history of the NNI, its evolution over the past decade, and the future prospects for this technology and its impact on society [6]. Additional information emphasizing aspects related to manufacturing at the nanoscale appears in a report issued by the National Nanotechnology Coordination Office [7].

- *The "1st generation" products (2000+): represented primarily by passive nanostructures.* The majority of products that are already commercialized fall into this category, where the nanoscale element (e.g., nanoparticle, nanoclay platelet, nanotube) is incorporated into a matrix material for coatings, films, and composites, or is part of a bulk nanostructured material. The processes for fabricating the target nanomaterials discussed in this book, as well as the products incorporating these nanoparticles represent the first generation of nanoproducts.
- *The "2nd generation" products (2005+): represented by active nanostructures.* In these structures, the nanoscale element is the functional structure, as in the case of nanospheres and nanostructured materials for drug delivery. The materials are functional in that they respond to some external stimuli such as pH or temperature to release the stored drug at a controlled rate. Other examples include sensors and actuators, transistors, and other electronics, where individual nanowires serve to provide the switching or amplifying mechanism.
- *The "3rd generation" products (2010+): represented by three-dimensional nanosystems and multi-scale architectures, expanding beyond the two-dimensional layer-by-layer approach currently used in microelectronics.* These systems will be manufactured using various directed self-assembly methods such as bio-assembly (e.g., using DNA and viruses as templates), electrical and chemical template-guided assembly.
- *The "4th generation" products (2015+): represented by truly heterogeneous molecular nanosystems.* In these products, multi-functionality and control of function will be achieved at the molecular level.

Common to all four generations of product development are three stages where exposure to nanomaterials is the most significant. In general, nanomaterials such as carbon nanotubes and silver nanoparticles can be relatively expensive, so companies will want to reduce waste as much as possible. Nevertheless, exposure and entry into the waste stream can occur: (1) during fabrication of the nanomaterial; (2) during storage and handling of the nanomaterial, including during incorporation of the nanomaterial into another material, structure, or device; and (3) during material removal or failure upon further processing or disposal of the product. Once the nanomaterial is incorporated into a bulk material (e.g., a carbon nanotube bonded within a polymer matrix), the concern is the same as that for the bulk material and is not related to the nanoscale dimensions or properties. Prior to embedment or in the case of release at end of life disposal, the unique properties of nanomaterials do have a very different effect. The most obvious case is that of worker exposure. With particles roughly 1/1000th the diameter of chopped glass fibers, the concern is that filtration and ventilation regulations are not effective. The behavior also is not monotonic with size. Some properties may actually make it easier to filter or collect any stray nanomaterials. For example, the Brownian motion of nanoparticles results in a more tortuous travel path that may make capture easier. Similarly, the high reactivity of the surface-dominated particles can lead to a greater ease of collection; for example, nanoparticles tend to agglomerate into much larger clusters, making them easier to detect and filter.

3.3.2 THE IMPACT OF "ENGINEERED" NANOMATERIALS

More than 10 years ago, as capabilities of measuring particles below 100 nanometers (nm) were developed, significant research focused on "ultrafine" particles resulting from vehicle emissions and combustion-related manufacturing processes such as welding. Since that initial research into nanoparticles as byproducts, interest in engineered nanoparticles has grown. The breadth of processes creating and utilizing nanoscale materials raises more challenges. Engineered nanomaterials are being created via multiple methods, for example, arc discharge, laser ablation, CVD, gas-phase synthesis, sol-gel synthesis, and high-energy ball milling. These processes can begin from the "bottom up," assembling nanomaterials from their components, for example by chemical synthesis or phase change processes. Other manufacturing methods begin with bulk materials, reducing their size via mass change processes to create nanomaterials from the "top down."

The bottom-up synthesis routes are, by far, the most widely used for nanoparticles. While engineered nanoparticles often are thought of as precursors or raw materials to be incorporated into higher value-added products via one of the five families of processes described previously in this chapter, the initial step of synthesizing nanoparticles most closely fits within the family of "phase change processes," which includes processes such as CVD. The use of top-down methods such as high-energy ball milling is limited to larger diameter particles with less stringent monodispersity and purity requirements. Ball milling is essentially a grinding process that would fit within the machining processes of the "mass change processes."

As with the other manufacturing processes, the process-structure-property interrelationships are significant. For example, the manufacturing process can affect the atomic structure of carbon nanotubes, which in turn affects many properties, such as the electrical conductivity (e.g., metallic vs. semiconducting), thermal conductivity, strength, and stiffness. One relatively coarse difference is the production of single-walled nanotubes (SWNTs) versus multi-walled nanotubes (MWNTs). Single-walled nanotubes have better conductivity and strength properties but are much less reactive and therefore more difficult to functionalize (i.e., to create compatibility with other materials for bonding). In general, the properties of nanoparticles are governed by process-induced factors such as the size and size distribution, degree of porosity, and surface reactivity. In synthesis processes, size and structure can be controlled through the use of catalyst particles, template materials (e.g., to control nucleation and precipitation behavior), and controlled-size droplets or aerosols.

The six nanomaterials that are the focus of this book — carbon black, carbon nanotubes, fullerenes (also known as C60 or buckyballs), nano silver, nano titanium dioxide, and nano zero-valent iron — can all be fabricated using many methods, and with the interest in nanomaterials, new methods are being discovered rapidly. A quick search in the U.S. Patent and Trademark Office database [8] brings up roughly 50 patents issued in the past two years with "nanoparticle" in the title. These patents include methods of making nanoparticles, modifying nanoparticles, and products incorporating nanoparticles. Table 3.1 provides a few examples of manufacturing techniques for the six target materials.

3.3.3   INTEGRATING NANOPARTICLES INTO NANOPRODUCTS

In some processes, the synthesis of the nanoparticle and subsequent deposition onto a substrate occurs in one continuous process. In others, however, the nanomaterial must be collected and stored until needed for later processing. Some earlier nanoparticle synthesis approaches resulted in the nanomaterial adhering to the walls of the reactor, requiring physical removal equivalent to "scraping the soot from the walls." Needless to say, such direct contact with the materials leads to worker exposure issues. Newer methods emphasize limiting human contact with the nanoparticles, partially for worker safety, but also for economic reasons: reducing contamination and increasing yield.

Once the nanomaterial is manufactured and sold as a raw material to multiple customers, the next stage of exposure is handling during incorporation into a product. Dispersion is often the key process in incorporating nanomaterials into bulk materials. This often involves some chemical modification of the surface to cause the nanomaterials to be less likely to agglomerate with each other and more likely to bond to the bulk material. Again, once in a solvent or suspension or melt, the nanomaterials are very unlikely to be inhaled, but dermal contact may still be a concern. Thus, the step of introducing the nanoparticles or nanotubes into the solution or melt is the potential hazard point. Beyond this point, the material remains in a closed environment (e.g., in a melt being mixed in a twin-screw extruder).

For future generations of products, the vision is that of three-dimensional multimaterial, directed self-assembly manufacturing processes. Simple two-dimensional examples include the organization of nanoparticles and other nanomaterials using conductive vs. nonconductive patterns (Figure 3.5) and the alignment of nanotubes in narrow trenches. In directed assembly, the material to be assembled (e.g., conductive polymer, nanoparticles, nanotubes) is exposed to the template. Then, with the help of some driving force such as an electric field, magnetic field, or chemical attraction, the nanomaterials assemble into a desired pattern over a large area within a short time. The benefit of these directed assembly processes is that the amount of handling will further decrease and the raw material is often in solution (e.g., not subject to inhalation). This is an advantage not only for repeatability, but also for worker exposure. The environmental question that then arises is the capture and reuse of

TABLE 3.1
Examples of Manufacturing Methods for Target Nanomaterials

Nanomaterial	Process Description	Ref.
Titanium dioxide	Minerals rutile (TiO_2) and ilmenite ($FeTiO_3$) are extracted from heavy mineral sands. Nanoscale particles can be manufactured by milling, but finer TiO_2 particles can be manufactured by a combination of chemical synthesis and milling: Prepare aqueous solution of $TiCl_4$ in solution with HCl, HPO_4. Vacuum-dry solution and spray-dry at 200–250°C to produce dry TiO_2. Calcinate at 600–900°C for 0.5–8 hours to produce crystalline nanostructure. Wash precipitate with C_2H_5OH, dry, and mill to nano-sized particles.	[9–11]
Zero-valent iron	The following processes are currently used in commercial production to manufacture nano zero-valent iron: React ferric chloride with sodium borohydride to create particles approximately 50 nm in diameter. React iron oxides (goethite and hematite) with hydrogen at 200–600°C to form particles approximately 70 nm in diameter containing Fe^0 and Fe_3O_4.	[12–14]
Silver	Silver is recovered from ore via smelting and electrolysis. Nanoscale particles can be created by several processes: Silver powder can be generated by atomizing molten silver (e.g., via a high-velocity gas jet) to create very small droplets that then solidify into powder form. Very fine silver particles are more commonly produced by chemical precipitation, e.g., from silver nitrate using a reducing agent (e.g., ascorbic acid). In a new variation on this process, the reaction occurs in a spinning disk processor (SDP); nanoparticles 5–200 nm form in a thin fluid film on the rotating disk surface. Nanometer-sized angular silver particles (1–100 nm) are produced when a high-power laser beam strikes a metallic block of silver immersed in a silver salt solution. Electrolysis of a silver electrode in deionized water produces colloidal silver containing both metallic silver particles (1–25 wt%) and silver ions (75–99 wt%).	[15–17]

TABLE 3.1 (CONTINUED)
Examples of Manufacturing Methods for Target Nanomaterials

Nanomaterial	Process Description	Ref.
Carbon black	Carbon black is produced from the incomplete combustion or thermal decomposition of hydrocarbons under controlled conditions. As the combustion products collide in the reactor, they form ever-larger particles by aggregation and agglomeration. Two methods produce most of the commercial carbon black: oil furnace and thermal black. The oil furnace process produces more than 95% of commercial carbon black. Preheated oil is atomized and partially combusted in a heated gas stream. The gas stream is quenched with water and carbon black is recovered on a bag filter. Recovered carbon black is mixed with water, then air-dried. The thermal black process, which entails the thermal decomposition of natural gas, accounts for most of the remaining production of carbon black.	[18, 19]
Carbon nanotubes	HiPco process of gas-phase chemical-vapor-deposition is currently in commercial use to manufacture single-walled carbon nanotubes (SWNTs): Introduce $Fe(CO)_5$ catalyst into injector flow via pressurized CO. Heat catalyst stream and mix with CO in graphite heater. $Fe(CO)_5$ decomposes to Fe clusters. Standard running conditions: 450 psi CO pressure, 1050°C. C atoms coat and dissolve around the Fe clusters, forming nanotubes. Running conditions maintained 24–72 hours. Gas flow carries SWNTs and Fe particles out of the reactor. SWNTs condense on filters. CO passes through NaOH absorbtion beds to remove CO_2 and H_2O, then is recycled.	[10, 11]
Fullerenes (C60 or buckyballs)	Fullerenes can be manufactured by several processes: Fires and lightning strikes naturally generate small amounts of fullerenes. Production in laminar benzene-oxygen-argon flame. Carbon arc discharged from graphite electrodes (Krätschmer-Huffman method).	[10, 11, 20]

excess solution and nanomaterial; this can, in a very rough sense, be considered similar to the problem of collection of cutting fluids in machining.

With respect to separation of the nanomaterial during further processing or disposal, an important question is the strength of the bonding. That is, how easily will the nanomaterials separate from a substrate and then potentially be freed into the environment? At the nanoscale, secondary bonds from van der Waals forces play a significant role. These bonds are, however, not as strong as chemical bonds (physisorption vs. chemisorption), although there are strong physisorption and weak chemisorption conditions that approach a middle ground. As with the current issues facing recycling of multi-material systems, the flip side to undesired nanomaterial liberation is the desire to easily separate materials for reuse upon disposal of the product.

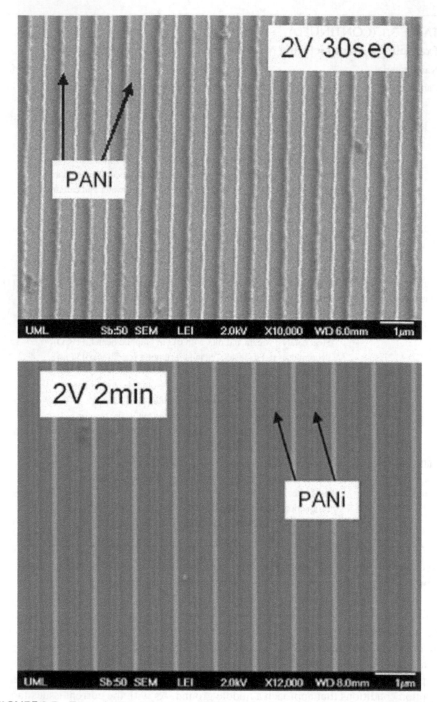

FIGURE 3.5 Example of template-directed assembly of a conductive polymer (doped poly-aniline, PANi) using 100-nm gold lines on a silicon wafer (assembly voltage and time is indicated on each image). (From Professor Joey Mead, Dr. Ming Wei, and Mr. Jia Shen, University of Massachusetts – Lowell, www.uml.edu/nano. With permission.)

3.4 SUMMARY

Understanding the environmental implications of any new technology is crucial to long-term sustainability. Unfortunately, such problems are complex, with many different points along the life cycle of fabrication, handling and integration, and disposal that must be addressed. Even within just one main process category, such as handling and integration of nanomaterials into nanoproducts, the breadth of different manufacturing processes and materials is vast, encompassing gas, liquid, and solid phases, as well as chemical, electrical, and mechanical deformation and assembly mechanisms. An all-inclusive answer to ensuring environmental safety and sustainability is not viable, but the remainder of this book addresses some of the existing nanomaterials that have shown relatively high volume commercial applicability. By understanding more about current nanomaterials and nanomanufacturing processes, the transfer of knowledge to yet-to-be-developed nanomaterials and processes will be invaluable.

REFERENCES

1. Nelson, D. 1980. *Frederick W. Taylor and the Rise of Scientific Management.* Madison, WI: The University of Wisconsin Press.
2. National Research Council. 1995. *Unit Manufacturing Processes: Issues and Opportunities in Research.* Washington, D.C.: National Academy Press.
3. NanoTex®. http://www.nano-tex.com. (Accessed November 30, 2007)
4. Okada, A. and A. Usuki. 2006. Twenty years of polymer-clay nanocomposites. *Macromolec, Mater. Eng.,* 291(12):1449–1476.
5. Sherman, L. 1999. Nanocomposites: a little goes a long way. *Plastics Technol.,* 45(6):52.
6. Roco, M.C. 2007. National Nanotechnology Initiative — Past, Present, and Future. In *Handbook of Nanoscience, Engineering, and Technology,* 2nd edition. Eds. W.A. Goddard III, D.W. Brenner, S.E. Lyshevski, and G.J. Iafrate. Boca Raton, FL: Taylor & Francis.
7. National Science and Technology Council, Subcommittee on Nanoscale Science, Engineering, and Technology. 2007. *Manufacturing at the Nanoscale: Report of the National Nanotechnology Initiative Workshops, 2002–2004.* Eds. J. Chen, H. Doumanidis, K. Lyons, J. Murday, and M.C. Roco. Washington: National Nanotechnology Coordination Office.
8. U.S. Patent and Trademark Office. http://www.uspto.gov/. (Accessed November 30, 2007)
9. Fisher, J. and T.A. Egerton. 2001. Titanium Compounds, Inorganic. In *Kirk-Othmer Encyclopedia of Chemical Technology.* New York: John Wiley & Sons, Inc.
10. Robichaud, C.O., D. Tanzil, U. Weilenmann, and M.R. Wiesner. 2005. Relative risk analysis of several manufactured nanomaterials: an insurance industry context. *Environ. Sci. Technol.,* 39(October):8985–8994.
11. Robichaud, C.O., D. Tanzil, U. Weilenmann, and M.R. Wiesner. 2005. Supporting information for relative risk analysis of several manufactured nanomaterials: an insurance industry context. *Environ. Sci. Technol.,* Published online October 4, 2005. http://pubs.acs.org/subscribe/journals/esthag/suppinfo/es0506509/es0506509si20050812_025756.pdf. (Accessed July 4, 2007)

12. Zhang, W.X. 2005. Nano-Scale Iron Particles: Synthesis, Characterization, and Applications. Meeting Summary: *U.S. EPA Workshop on Nanotechnology for Site Remediation*. Washington, D.C. 20–21 October. http://www.frtr.gov/nano.

13. Vance, D. 2005. Evaluation of the control of reactivity and longevity of nano-scale colloids by the method of colloid manufacture. *Meeting Summary: U.S. EPA Workshop on Nanotechnology for Site Remediation*. Washington, D.C. 20–21 October. http://www.frtr.gov/nano.

14. Wiesner, M.R., G.V. Lowry, P. Alvarez, D. Dionysiou, and P. Biswas. 2006. Assessing the risks of manufactured nanomaterials. *Environ . Sci. Technol.*, 40(14):4336–4345.

15. Etris, A.F. 2001. Silver and silver alloys. In *Kirk-Othmer Encyclopedia of Chemical Technology,* 4:761–803. New York: John Wiley & Sons, Inc.

16. Key, F.S. and G. Maas. 2001. Ions, Atoms and Charged Particles. http://www.silver-colloids.com/papers/IonsAtoms&ChargedParticles. (Accessed October 6, 2007)

17. Iyer, K.S., C.L. Raston, and M. Saunders. 2007. Transforming nano-science to nano-technology: manipulating the size, shape, surface morphology, agglomeration, phases and defects of silver nano-particles under continuous flow conditions. *NSTI-Nanotech 2007*, p. 4.

18. Wang, M.-J., C.A. Gray, S.A. Resnek, K. Mahmud, and Y. Kutsovsky. 2003. Carbon black. In *Kirk-Othmer Encyclopedia of Chemical Technology* 4:761–803. New York: John Wiley & Sons, Inc.

19. International Carbon Black Association. 2004. Carbon Black User's Guide. http://carbon-black.org. (Accessed August 30, 2007)

20. Holister, P., C. Roman Vas, and T. Harper. 2003. Fullerenes: Technology White Papers nr. 7. Cientifica, Ltd. (October). http://www.cientifica.com.

4 Developing Environmental Regulations Pertinent to Nanotechnology

Lynn L. Bergeson
Bergeson & Campbell, P.C.

CONTENTS

Many governments around the world are deeply committed to promoting the responsible development of nanotechnology and are engaged in a wide variety of nanotechnology initiatives. These initiatives are expressed in multiple venues — research and development projects, policy pronouncements, and various regulatory initiatives across federal agencies and departments. This chapter addresses key nanotechnology regulatory initiatives underway at the U.S. Environmental Protection Agency (EPA), and regulatory developments at several other federal agencies and departments in the United States. It also provides an overview of regulatory programs in the European Union and Canada, and initiatives by the Organization for Economic Cooperation and Development.

4.1 THE TOXIC SUBSTANCES CONTROL ACT (TSCA)

4.1.1 TSCA Statutory and Regulatory Background

The federal law implemented and enforced by the EPA most often cited in connection with regulating nanoscale substances that are intentionally manipulated by human activity, and not on naturally occurring nanoscale particles (volcanic ash) or incidental nanoscale materials (combustion byproducts), is the Toxic Substances Control Act (TSCA) [1]. TSCA regulates new and existing chemical substances and provides a regulatory framework to address chemicals throughout their production, use, and disposal.*

Enacted by the U.S. Congress in 1976 to protect human health and the environment from the effects of exposure to potentially harmful chemical substances and mixtures, TSCA is the federal statute that authorizes the EPA to regulate engineered nanoscale materials that are chemical substances. TSCA is interpreted broadly and is directed toward regulating "chemical substances" [2] through their manufacture, use, and disposal. The term "chemical substance" means "any organic or inorganic substance of a particular molecular identity, including — any combination of such substances occurring in whole or in part as a result of a chemical reaction or occurring in nature, and any element or uncombined radical."**

TSCA applies broadly to "any person" who manufactures, processes, distributes in commerce, uses, or disposes of a chemical substance. TSCA requirements fall most heavily on chemical manufacturers. For TSCA purposes, "manufacture"

---

* Other articles on this subject include: Bergeson, L.L. and J.E. Plamondon. 2007. TSCA and Engineered Nanoscale Substances. Nanotechnol., Law & Bus., 4(1): 51; Bergeson, L. and B. Auerbach. 2004. Reading the Small Print. Env't Forum, 30–41; Breggin, L.K. 2005. Securing the Promise of Nanotechnology: Is the U.S. Environmental Law up to the Job? A Dialogue. Envtl. Law Inst. (Woodrow Wilson Int. Ctr. for Scholars Project on Emerging Nanotech.). October. Available at http://www. elistore.org/reports detail.asp?ID=11116; American Bar Association (ABA). 2006. Regulation of Nanoscale Materials under the Toxic Substances Control Act, Section of Environment, Energy, and Resources (SEER). June. Available at http://www.abanet.org/environ/nanotech/pdf/TSCA.pdf.

** TSCA § 3(2)(A), 15 U.S.C. § 2602(2)(A) (2007); See also EPA 40 C.F.R. §§ 710.3(d), 720.3(e) (2007). TSCA Section 3(2)(B) excludes from the definition of "chemical substance" mixtures, pesticides, tobacco and tobacco products, certain nuclear materials, firearms and ammunition, and foods, food additives, drugs, cosmetics, and devices. 15 U.S.C. § 2602(2)(B) (2007); see also EPA 40 C.F.R. §§ 710.3(d), 720.3(e) (2007). All of these categories, with the exception of mixtures, are regulated under other federal laws. The TSCA defines a "mixture" as "any combination of two or more chemical substances if the combination does not occur in nature and is not, in whole or in part, the result of a chemical reaction." Also included within the definition is any chemical substance that is the result of a chemical reaction, but that could have been manufactured for commercial purposes without a reaction. TSCA § 3(8), 15 U.S.C. § 2602(8) (2007); see also EPA 40 C.F.R. §§ 710.3(d), 720.3(u) (2007). In addition to these statutory exclusions, the EPA's regulations exclude "articles" and other types of substances (e.g., certain impurities and by-products) for purposes of various TSCA provisions. See, e.g., EPA 40 C.F.R. §§ 704.5, 710.4(d), 720.30 (2007).

includes importation.* This definition brings importers of chemical substances within TSCA's jurisdictional reach, even though actual chemical manufacturing activities occur outside of the United States.

TSCA governs both the manufacture of "new" chemical substances and regulates uses of "existing" chemical substances that the EPA has determined to be "significant new" uses [3]. TSCA Section 8(b)(1) directs the EPA to "compile, keep current, and publish a list of each chemical substance ... manufactured or processed in the United States." [4] The majority of the chemicals included on the TSCA Inventory are substances that were in commerce prior to December 1979, and are so listed because entities included them on the Inventory when it was first published on June 1, 1979 [5]. Under TSCA, these substances are considered "existing" chemical substances by virtue of their listing on the Inventory [6, 7]. The Inventory is updated with chemical substances that have been added since the original Inventory was issued in 1979, including those chemicals the EPA has more recently reviewed and approved as "new" chemicals subject to the premanufacture notification (PMN) provisions under TSCA Section 5. Thus, under TSCA, a chemical substance is considered either an "existing" chemical substance (because it is included on the Inventory) for TSCA purposes, or a "new" chemical substance (because it is not and must be approved by the EPA prior to manufacture). For engineered nanoscale materials, the distinction is particularly significant.** A "significant new" use of a chemical substance already listed on the TSCA Inventory is treated much like a new chemical substance, and the new use is subject to EPA review in much the same way that the EPA reviews a new chemical.

To ensure compliance with TSCA, prior to the commercial manufacture of a chemical substance for a non-exempt purpose, the manufacturer must first determine its TSCA Inventory status. There are two Inventories: (1) the Public Inventory and (2) the Confidential Inventory. If a search of the Public Inventory (which is included on a publicly available, searchable database) [8] does not yield a listing, the next step is to determine whether the substance is included on the Confidential Inventory. If the identity of a chemical substance has been claimed as a trade secret, or otherwise it is not listed on the Public Inventory, it may be listed on the TSCA Confidential Inventory. To determine if it is listed, a *bona fide* intent (BFI) request must be submitted to the EPA so that the EPA can search the Confidential Inventory [9].

If a chemical substance is not listed on either portion of the TSCA Inventory, manufacturers must submit a PMN for any chemical substance to be manufactured and that is not eligible for a PMN exemption. The PMN form itself is straightforward and seeks information only on the submitter's identity, and the chemical substance's identity, production volume, uses, exposures, and environmental fate [10].

---

* TSCA § 3(7), 15 U.S.C. § 2602(7) (2007). Under the implementing regulations for TSCA Sections 5 and 8, "manufacture" is defined to mean "to manufacture, produce, or import for commercial purposes," which in turn is defined to mean "to manufacture, produce, or import with the purpose of obtaining an immediate or eventual commercial advantage." See, e.g., EPA 40 C.F.R. § 710.3(d) (2007).

** The PMN regulations are at EPA 40 C.F.R. pt. 720 (2007), and PMN exemptions are at EPA 40 C.F.R. pt. 723 (2007). Existing chemical substances already listed on the TSCA Inventory may be subject to a Significant New Use Rule (SNUR), which also is authorized under TSCA Section 5 and EPA 40 C.F.R. pt. 721, subpart E, and is discussed below.

TSCA does not require the PMN submitter to test a new chemical substance before submitting a PMN. Health and safety data relating to a new chemical substance's health or environmental effects that are in a submitter's possession or control, however, must be submitted along with the PMN to the extent it "is known to or reasonably ascertainable by" the submitter.* The period for the EPA review of a PMN is 90 days, unless extended by the EPA for up to an additional 90 days.**

There are several exemptions from the PMN requirements, some of which are relevant to engineered nanoscale materials that are chemical substances. The TSCA exemptions fall into one of two categories: (1) self-executing, and (2) those that require EPA approval. Exemptions are considered "self-executing" because they do not require prior EPA approval; and once a manufacturer determines that one of the self-executing exemptions applies, the new chemical substance can be manufactured in the United States without first submitting a PMN. However, the entity must comply with certain recordkeeping and/or other requirements for the particular exemption to apply. Self-executing exemptions include the exemption for chemical substances having no separate commercial purpose, the polymer exemption, and the research and development (R&D) exemption.

Other exemptions from PMN requirements require EPA prior approval. In these instances, a manufacturer must submit, and the EPA must approve, an exemption application before a company can commence manufacture of the new chemical substance, subject to compliance with any associated recordkeeping and/or other requirements that may apply. These exemptions are for low volume (LVE), low release and low exposure (LoREX), and the test marketing exemption (TME).

The self-executing R&D exemption is particularly important to the emerging nanotechnology industry [11, 12]. To qualify as an R&D substance, the chemical substance must be manufactured or imported only in "small quantities" for purposes of scientific experimentation or analysis, or for chemical research on or analysis of such substance or another substance, including such research or analysis for the development of a product [13]. The term "small quantities" is not defined quantitatively, but qualitatively, as those "that are not greater than reasonably necessary" for R&D purposes [13, 14]. Substances that satisfy the criteria for an R&D substance must be used by or under the supervision of a "technology qualified individual" (TQI), who is tasked with ensuring compliance with volume, prescribed uses, labeling, handling and distribution, disposal, and recordkeeping requirements.

Two other exemptions that are relevant to emerging nanotechnology industries — the LVE and LoREX exemptions — are not self-executing and require explicit EPA approval. These exemptions require prior EPA review and approval, and the process for obtaining EPA approval can be time consuming and resource intensive.

---

* See EPA 40 C.F.R. §§ 720.40(d), 720.50 (2007). The phrase "known to or reasonably ascertainable by" is defined at EPA 40 C.F.R. § 720.3(p) (2007).
**TSCA § 5(a), (c), 15 U.S.C. § 2604(a), (c) (2006); EPA 40 C.F.R. § 720.75. The review period can be extended repeatedly.

Notice must be submitted at least 30 days before manufacture begins, triggering a 30-day period for EPA review and action.*

Eligibility for an LVE is based on the manufacture of a new chemical in quantities of 10,000 kilograms (kg) or less per year.** Eligibility for a LoREX is based on meeting several regulatory criteria throughout the processes of manufacturing, processing, distribution, use, and disposal of the chemical substance. These include, for consumers and the general population, no dermal or inhalation exposure and no drinking water exposure greater than 1 milligram (mg) per year. For workers, there can be no dermal or inhalation exposure; there can be no releases to ambient surface water in concentrations above 1 part per billion (ppb); no releases to the ambient air from incineration in excess of 1 microgram per cubic meter (1 $\mu g/m^3$); and no releases to groundwater, land, or a landfill unless it is demonstrated that there is negligible groundwater migration potential [15]. Once the EPA notifies the applicant that an exemption has been granted, or if the review period expires without notice from the EPA, manufacture or import of the chemical substance can commence, consistent with the terms of the exemption.

4.1.2 EPA OPPT Nanotechnology Initiatives

The EPA's Office of Pollution Prevention and Toxics (OPPT), the program office tasked with implementing TSCA, has been very active over the past several years in the nanotechnology area. Several initiatives are relevant, each of which is described below.

4.1.2.1 Nanoscale Materials Stewardship Program

The OPPT announced in 2005 its interest in considering how best to obtain much-needed data and information on existing engineered nanoscale materials, and convened, in June 2005, a public meeting to discuss various options [16]. The discussion at the public meeting yielded a consensus that a voluntary program designed to obtain existing and new information, and new data on engineered nanoscale substances has significant value.

Shortly thereafter, the EPA decided to create an Interim Ad Hoc Work Group on Nanoscale Materials (Work Group) as part of the National Pollution Prevention and Toxics Advisory Committee (NPPTAC), a federal advisory group tasked with advising the OPPT on TSCA and related pollution prevention matters. The Work Group was formed to provide input to the NPPTAC on the need for, and design of, a voluntary program for reporting information pertaining to existing chemicals that are engineered nanoscale materials, and the information needed to inform adequately the conduct of such a program.

---

* EPA 40 C.F.R. § 723.50(a)(2), (g) (2007). This review period can be suspended to allow the EPA a longer review period. The EPA approved the first LoREX for what is believed to be a single wall carbon nanotube in October 2005. The review and approval period was 13 months. See, e.g., TSCA § 5(a), (c), 15 U.S.C. § 2604(a), (c) (2006); EPA 40 C.F.R. § 720.75. The review period can be extended repeatedly..
** See EPA 40 C.F.R. § 723.50(a), (c) (2007). One kilogram is equivalent to 2.2 pounds.

On November 22, 2005, the NPPTAC issued its *Overview Document on Nanoscale Materials*, which outlines a framework for an EPA approach to a voluntary program for engineered nanoscale materials, a complementary approach to new chemical nanoscale requirements under TSCA, and various other relevant issues pertinent to engineered nanoscale materials that are chemical substances [17, 18]. The voluntary program was named the Nanoscale Materials Stewardship Program (NMSP). The Overview Document provides that the "overall goal of EPA's program regarding engineered nanoscale materials should focus on addressing the potential risks of such materials to human health and the environment, thereby giving the public reasonable assurances of safety concerning such materials."

Inclusion of the expression "reasonable assurances of safety" was questioned by some NPPTAC members on the grounds that it could be interpreted as suggesting a standard different from the "may present an unreasonable risk" standard set forth in TSCA's statutory language. The NPPTAC ultimately agreed that the "assurances of safety" language as an "overall goal" of the NMSP was not reasonably likely to supplant the TSCA legal standard, and that it fairly articulated the overall goal of the EPA's program regarding engineered nanoscale materials.

Scope of the Program: According to the Overview Document, the voluntary program is intended to encompass engineered nanoscale materials now in or "soon to enter" commerce. "Soon to enter" was defined as "applying to pre-commercial new and existing chemical engineered nanoscale materials for which there is clear commercial intent on the part of the developer, excluding such materials that are only at the research stage, or for which commercial application is more speculative or uncertain."

Elements of the NPPTAC Program: The Work Group expressed its view that program participants should be offered the choice of participating in a "Basic" Program or in a more "In-Depth" Program that included, in addition to all the elements of the Basic program, the commitment to generate and report more in-depth information, and implement more in-depth risk management practices.

Both of the proposed programs — Basic and In-Depth — are voluntary, and participation in either would, according to the NPPTAC, offer benefits for those willing to provide information and agree to implement appropriate risk management practices. Under the NMSP, participants would volunteer one or more specific engineered nanoscale materials that they are developing, producing, processing, or using, but need not necessarily volunteer all of their materials. Information provided by participants relevant to understanding and addressing the potential risks of engineered nanoscale materials will be made publicly accessible, limited as appropriate by protections applicable to confidential business information (CBI) as described under TSCA.

Basic Program Participation: Participation in the Basic Program of the NMSP would consist of the following three sets of activities for each volunteered engineered nanoscale material:

1. Reporting existing (hereinafter meaning all information possessed by the submitter) material characterization information on the material, as well as existing information characterizing hazard, use, and exposure potential, and risk management practices
2. Filling in gaps on basic information about material characteristics only
3. Implementing basic risk management practices

A core element of the voluntary program envisioned by the Work Group is reporting existing information, which refers to all information in the possession of the submitting company. The information reported on each volunteered nanoscale material would include the following:

• Existing material characterization information on engineered nanoscale materials
• Existing information on hazards (i.e., environmental fate and toxicity studies)
• Existing information about use and exposure potential
• Existing information about risk management and other protective measures implemented now or available to be applied to engineered nanoscale materials, and to products and wastes containing such materials*

If elements of a baseline set of material characterization information are missing, voluntary program participants are expected to generate the missing information. The baseline would consist of the following basic material characterization information: chemical composition (including impurities), aggregation/agglomeration state, physical form, concentration, size distribution and/or surface area, and solubility. It is believed that most producers, processors, users, and researchers already have this type of information about materials characteristics, and that this commitment would result in only a minimal additional burden.

Participation in the Basic Program would include a risk management component that consists of a participant's agreement to implement basic risk management practices or other environmental or occupational health protection controls (e.g., worker training, hazard communication, material safety data sheets, use of available engineering controls, provision of personal protective equipment [PPE], product labeling, customer training, waste management practices, etc.). Participants also are expected to describe their experience in implementing, and their degree of satisfaction with, Basic Program risk management practices.

In-Depth Program Participation. The In-Depth Program is for organizations, or consortia of organizations and/or entities, that are interested in participating beyond the Basic Program. Participants would agree to generate new information about the hazard and risks (including reduction of risk) of a particular engineered nanoscale material, as well as identifying, implementing, and expanding, as needed, risk

---

* In this regard, the EPA convened a second peer consultation on September 6–7, 2007, to discuss nano-materials characterization.

management measures appropriate for a given life-cycle phase of such substance. According to the Overview Document:

> The In-Depth Program would be expected to focus on a more limited number of engineered nanoscale materials, generating and reporting more in-depth information as identified by EPA as necessary to allow the Agency to conduct a full risk assessment of the identified materials and associated uses. For each volunteered material, producers, processors, users, and researchers and/or consortia of such entities would submit Basic Program information and would concurrently begin to generate the additional, more in-depth information, although it is expected that it will take longer to generate the new information. In-depth information on the engineered nanoscale materials would be submitted on a prescribed set of elements, developed by EPA in advance of program launch, on material characterization, human health hazard, environmental hazard, and release and exposure. The information would be generated with an aim to avoid redundancy and ensure efficient use of resources [17].

Under the In-Depth Program, volunteers also would agree to work to extend application of protective risk management practices identified by the EPA along their supply chains, and to conduct monitoring of workplaces, environmental releases, and worker health.

An aspect of the voluntary program that attracted considerable attention was program evaluation. The program is intended to be time limited, and it is expected that the EPA will determine a point in time at which it will conduct a full-scale program evaluation to assess at least the following:

- Degree to which the program is meeting its goals
- Rate of participation
- Amount and quality of the information generated by the program participants
- Adequacy and potential effectiveness of existing risk management practices
- Lessons and conclusions that can be drawn from the program experience

NPPTAC members, and especially the Interim Ad Hoc Work Group members, expressed keen interest in ensuring that the program did not simply get off the ground, but also that it meet the EPA's intended goals within a reasonable period of time.

On October 18, 2006, the Office of Prevention, Pesticides, and Toxic Substances (OPPTS) Assistant Administrator, Jim Gulliford, sent a letter to stakeholders formally announcing the development of the NMSP and inviting stakeholder participation in it [19]. According to the letter, the EPA's goal "is to implement TSCA in a way that enables responsible development of nanotechnology and realizes its potential environmental benefits, while applying sound science to assess and, where appropriate, manage potential risks to human health and the environment presented by nanoscale materials" [19]. The letter explained:

Over the coming months, we will be announcing a variety of opportunities for public input regarding our program to address nanoscale materials including: (1) public
scientific peer consultations to discuss risk management practices and characterization
for nanoscale materials; (2) an overall framework document describing the TSCA program for nanoscale materials; (3) a document on distinguishing the TSCA Inventory
status of 'new' versus 'existing' chemical nanoscale materials; (4) a concept paper
describing EPA's thinking for the Stewardship Program, as well as an Information Collection Request to collect data under the Stewardship Program; (5) workshops examining the pollution prevention opportunities for nanoscale materials; and (6) a public
meeting to discuss these documents and program elements [19].

On July 12, 2007, the EPA issued a "concept paper" on the NMSP; convened a
public stakeholder meeting on August 2, 2007; and requested public input on the elements of the program [20]. Each of these developments is discussed below.

The NMSP Concept Paper describes the EPA's general approach, issues, and considerations for the NMSP, and is intended to serve as a starting point for continuing
work with stakeholders on the detailed design of the NMSP. The EPA developed the
NMSP Concept Paper and its accompanying annexes "to outline [the EPA's] initial
thinking on the design and development" of the NMSP, which will "complement and
support [the EPA's] new and existing chemical efforts on nanoscale materials" and
"help address some of the issues identified in the EPA's Nanotechnology White Paper."
[21] The EPA states that the NMSP has the following specific objectives [21]:

*   Help the EPA assemble existing data and information from manufacturers
    and processors of existing chemical nanoscale materials
*   Identify and encourage the use of risk management practices in developing
    and commercializing nanoscale materials
*   Encourage the development of test data needed to provide a firmer scientific
    foundation for future work and regulatory/policy decisions
*   Encourage responsible development

The NMSP will include, but not be limited to, engineered nanoscale materials
manufactured or imported for commercial purposes within the meaning of 40 C.F.R.
Section 720.3(r). Importantly, the EPA explains that participation in the NMSP
"would not relieve or replace any requirements under TSCA that a manufacturer,
importer, or processor of nanoscale materials may otherwise have" [21].

Annex A of the NMSP Concept Paper ("Description of Nanoscale Materials
for Reporting") contains "clarifications and descriptions" of various key terms used
throughout the Concept Paper, including "engineered," "nanoscale," "engineered
nanoscale material," and "nanotechnology."

With respect to participation in the NMSP, the EPA foresees involvement by
persons or entities that do or intend to do any of the following, with the corresponding intent to offer a commercially available product: manufacture or import
engineered nanoscale materials; physically or chemically modify an engineered
nanoscale material; physically or chemically modify a non-nanoscale material to
create an engineered nanoscale material; or use engineered nanoscale materials in
the manufacture of a product [21]. Both "new" and "existing" (for purposes of TSCA

Section 5) engineered nanoscale materials can be included in the NMSP. Annex A also provides examples of materials that the EPA believes would and would not be appropriate for inclusion in the program.

Consistent with the NPPTAC Interim Ad Hoc Work Group on Nanoscale Materials' recommendations, the EPA is considering a two-part NMSP: (1) a "Basic" Program that would request the reporting of "all known or reasonably ascertainable information regarding specific nanoscale materials," and (2) an "In-Depth" Program in which additional data would be developed and submitted to the EPA over a longer timeframe [21]. Annex B (Data Elements) delineates the types of data that participants in the Basic Program would be expected to report. Submitters would be encouraged, but not required, to submit their data through a data submission form that the EPA has prepared.* Data claimed as CBI will be protected "in the same manner as CBI submitted under TSCA in accordance with procedures in 40 CFR parts 2 and 720" [21], and the EPA encourages NMSP participants both "to give careful consideration to what they will and will not claim [as] CBI" and "to make as much data as possible available to the public" [21].

As part of the Basic Program, NMSP participants would agree to implement a risk management program, as well as "agree to consider information provided by EPA that is relevant to [nanoscale material] risk management ... and to provide information about the risk management practices and other aspects of their risk management program that are relevant to nanoscale materials" [21].

The In-Depth Program would be informed by the Basic Program's results, and would involve a subset of the information reported under the Basic Program "in a greater amount of detail" [21]. The EPA states that "[i]n-depth data development would likely apply to a smaller set of representative nanoscale materials designated for further evaluation by mutual agreement of EPA and participants, with input from stakeholders" [21].

The EPA will use the data from the NMSP "to gain an understanding of which nanoscale materials are produced, in what quantities, how they are used, and the data that is available for such materials" [21]. The data will assist EPA scientists in making human health and environmental risk determinations, and may be used to "[i]dentify the data that are missing to conduct an informed risk assessment of a specific nanoscale material" and "[i]dentify nanoscale materials or categories of nanoscale materials that may not warrant future concerns or actions, or should otherwise be treated as a lower priority for further consideration" [21]. Significantly, the EPA explains that if data submitted by an NMSP participant "indicates that the participant is manufacturing a nanoscale material that is reportable under [TSCA] Section 5 ... as a new chemical substance, EPA will immediately inform the participant of that situation and the applicable TSCA requirements" [21].

Roughly a year after commencing the Basic Program, the EPA will publish an interim report summarizing "the types of data available, the reasons some data were reported as not being available, additional data that would be needed for a better risk assessment and any activities for which data are being used." Two years after the

---

* The draft submission form, which is based on the EPA's Premanufacture Notice (PMN) form (i.e., EPA Form 7710-25), is available at http://www.epa.gov/opptintr/nano/nmsp-icr-reportingform.pdf.

launch of the NMSP, the EPA will issue a more detailed evaluation of the program and simultaneously "determine the future direction of the basic reporting phase as well as in-depth data development" [21].

The EPA stated that it will work collaboratively with other federal agencies and stakeholders to develop further and implement the NMSP. Although dependent on the outcome of this development process, the EPA envisions that the components of the NMSP could include:

- Assembling existing data and information from manufacturers and processors of existing chemical nanoscale materials
- Encouraging the development of test data needed to provide a firmer scientific foundation for future work and regulatory/policy decisions
- Identifying and encouraging the use of a basic set of risk management practices in developing and commercializing nanoscale materials

The EPA will use the data from the NMSP to gain an understanding of which nanoscale materials are produced, in what quantities, how they are used, and the data available for such materials. EPA scientists will use data collected through the NMSP, where appropriate, to aid in determining how and whether certain nanoscale materials or categories of nanoscale materials may present risks to human health and the environment. The EPA requests comment on specific issues [21].

The draft Information Collection Request (ICR) on which the EPA requested comment on July 12, 2007, covers the information collection-related activities related to the NMSP and the estimated paperwork burdens associated with those activities. The EPA solicited comment on specific aspects of the proposed information collection for the voluntary NMSP.

In its draft TSCA Inventory "current thinking" document, the OPPT describes its "general approach" to determining whether a nanoscale substance meeting the definition of a chemical substance is "new" for TSCA purposes based on EPA guidance issued on July 12, 2007.* In the guidance document, the EPA reaffirms its policy not to use particle size to distinguish, for Inventory purposes, substances that are known to have the same molecular identity. The EPA states that molecular identity is "based on such structural and compositional features," including the types and number of atoms in the molecule, the types and number of chemical bonds, the connectivity of the atoms in the molecule, and the spatial arrangement of the atoms within a molecule. Chemical substances that "differ" in any of these structural or compositional features, according to the EPA, have different molecular identities.

Importantly, the EPA states that substances have different molecular identities when they: have different molecular formulas, have the same molecular formulas but different atom connectivities, have the same molecular formulas and atom connectivities but different spatial arrangements of atoms, have the same types of atoms but different crystal lattices, are different allotropes of the same element, or have different isotopes of the same elements.

---

* The document is available at http://www.epa.gov/oppt/nano/nmspfr.htm.

In the "current thinking" document, the EPA encourages nanoscale material manufacturers to arrange a pre-notice consultation with the EPA to address these issues, or to submit a *bona fide* intent to manufacture submission. The EPA also notes that it may need additional information, including data, to determine whether a material requires new chemical notification.

Ultimately, the EPA hopes the NMSP will be more robust than the Voluntary Reporting Scheme for Engineered Nanoscale Materials launched in September 2006 under the auspices of the U.K. Department for Environment, Food, and Rural Affairs (Defra). As of this writing, only nine entities have volunteered for the program. The objective of the U.K. program is "to develop a better understanding of the properties and characteristics of different engineered nanoscale materials, so enabling potential hazard, exposure and risk to be considered" in the U.K. government's effort "to develop appropriate controls in respect of any risks to the environment and human health from free engineered nanoscale materials" [22, 23].

4.1.2.2   Nanotechnology White Paper

The EPA's Science Policy Council (SPC) issued, in December 2005, its draft *Nanotechnology White Paper*. The White Paper describes issues the EPA must address to ensure that "society accrues the important benefits to environmental protection that nanotechnology may offer, as well as to understand better any potential risks from exposure to nanomaterials in the environment" [24]. The EPA convened an expert peer review meeting in Washington, D.C., on April 19–20, 2006, to conduct an independent expert external peer review of the White Paper [25]. The SPC approved the final report on September 25, 2006, and the EPA issued the final White Paper on February 15, 2007 [26].

The White Paper includes a discussion of the potential environmental benefits of nanotechnology, an overview of existing information on nanomaterials regarding components needed to conduct a risk assessment, a section on responsible development and the EPA's statutory mandates, and a review of research needs for both environmental applications and implications of nanotechnology. To help the EPA focus on priorities for the near term, the White Paper also provides recommendations for addressing science issues and research needs, including prioritized research needs within most risk assessment topic areas (e.g., human health effects research, fate and transport research). The EPA's Nanotechnology Research Framework, which is appended to the White Paper in Appendix C, outlines how the EPA intends to focus its own research program "to provide key information on potential environmental impacts from human or ecological exposure to nanomaterials in a manner that complements other federal, academic, and private-sector research activities."

Key White Paper recommendations include:

- *Environmental Applications Research.* The EPA should continue to undertake, collaborate on, and support research to understand and apply information regarding environmental applications of nanomaterials.

- *Risk Assessment Research.* The EPA should continue to undertake, collaborate on, and support research to understand and apply information regarding nanomaterials':
  - Chemical and physical identification and characterization
  - Environmental fate
  - Environmental detection and analysis
    - Potential releases and human exposures
    - Human health effects assessment
    - Ecological effects assessment

To ensure that research best supports EPA decision making, the EPA should conduct case studies to identify unique risk assessment considerations for nanomaterials.

- *Pollution Prevention, Stewardship, and Sustainability.* The EPA should engage resources and expertise to encourage, support, and develop approaches that promote pollution prevention; sustainable resource use; and good product stewardship in the production, use, and end-of-life management of nanomaterials. Additionally, the EPA should draw on new, next-generation nanotechnologies to identify ways to support environmentally beneficial approaches such as green energy, green design, green chemistry, and green manufacturing.
- *Collaboration and Leadership.* The EPA should continue and expand its collaborations regarding nanomaterial applications and potential human health and environmental implications. More specifically, the White Paper recommends the following actions:
  - The EPA's Office of Research and Development (ORD) should collaborate with other groups on research into the environmental applications and implications of nanotechnology. The ORD's laboratories should put a special emphasis on establishing Cooperative Research and Development Agreements (CRADAs) to leverage non-federal resources to develop environmental applications of nanotechnology. (CRADAs are established between the EPA and research partners to leverage personnel, equipment, services, and expertise for a specific research project.)
  - The EPA should collaborate with other countries (e.g., through the Organization for Economic Cooperation and Development) on research on potential human health and environmental impacts of nanotechnology.
  - The EPA's Office of Congressional and Intergovernmental Relations should lead efforts to investigate the opportunities for collaboration with and through state and local government economic development, environmental, and public health officials and organizations.
  - The EPA's Office of Public Affairs and program offices, as appropriate, should lead an EPA effort to implement the communication strategy for nanotechnology.
  - The Office of Policy, Economics, and Innovation's Small Business Ombudsman should engage in information exchange with small businesses, which comprise a large percentage of U.S. nanomaterial producers.

- *Intra-Agency Workgroup.* The EPA should convene a standing intra-agency group to foster information sharing on nanotechnology science and policy issues.
- *Training.* The EPA should continue and expand its nanotechnology training activities for scientists and managers [24].

According to the White Paper, as new generations of nanomaterials evolve, so will new and possibly unforeseen environmental issues. The White Paper states that it will be crucial that the EPA's "approaches to leveraging the benefits and assessing the impacts of nanomaterials continue to evolve in parallel with the expansion of and advances in these new technologies" [24].

4.1.2.3   TSCA PMN Decision Logic

The EPA's OPPT has developed and continues to reference a decision logic that OPPT staff applies in assessing engineered nanoscale materials that are chemical substances, when those nanoscale materials are reported to the EPA either under the PMN provision of TSCA, or as exemption applications therefrom. Use of the decision logic is resulting in EPA's identification of specific areas of inquiry unique to engineered nanoscale materials that are chemical substances. Primary among these areas are potential routes of exposure to workers and potential environmental releases of these materials. The EPA is assessing the adequacy of PPE to prevent potential exposures to engineered nanoscale materials during the manufacturing, processing, and/or distribution and use of these materials. The EPA's decision logic is believed to distinguish between "true" engineered nanoscale materials, meaning those that meet the criteria set out by the National Nanotechnology Initiative (NNI), and those materials that fall within the size range of 1 to 100 nm but are not specifically engineered with the intent to enable novel, size-dependent properties. According to published sources, the EPA has, as of August 2006, reviewed 15 new chemicals that were deemed to fall within the "nanoscale" size range, only one of which, siloxane-coated alumina nanoparticles, the EPA believed possessed properties deemed "unique."* According to EPA sources, the siloxane-coated alumina nanoparticles will have non-dispersive uses as an additive to other chemical substances.

4.2   THE FEDERAL INSECTICIDE, FUNGICIDE, AND RODENTICIDE ACT (FIFRA)

4.2.1   FIFRA Statutory and Regulatory Background

The EPA recognizes that there are many promising agricultural and antimicrobial applications of nanotechnologies and nanoscale substances. Nanosensors offer the promise of real-time pathogen detection/location reporting using nanotechnologies

---

* Pat Phibbs-Rizzuto, EPA Reviews 15 New Nanoscale Chemicals, but Finds Only One with Unique Properties, 158 Daily Env't Rep. (BNA) A-7 (Aug. 16, 2006). On August 14, 2006, EPA issued a notice acknowledging receipt of a notice of commencement of manufacture or import of siloxane-coated alumina nanoparticles pursuant to TSCA Section 5. 71 Fed. Reg., 46475, 46480 (Aug. 14, 2006).

in micro-electrochemical system technology. Increased biological efficiency could result in diminished amounts of pesticides being applied. Similarly, nanodevices used for "smart" treatment delivery systems hold promise.

Pesticide product registration is the central mechanism for regulating pesticide sales and use in the United States. Under the Federal Insecticide, Fungicide, and Rodenticide Act (FIFRA) [27], the EPA makes an individual registration determination for each pesticide product based on a separate application for registration. To issue a registration, the EPA must determine, among other findings, that the product will function without "unreasonable adverse effects on the environment," and when used in accordance with widespread and commonly recognized practice, will not generally cause unreasonable adverse effects on the environment.

Pesticide registrations include extensive data requirements for the EPA to evaluate the environmental effects, human health effects, and safety of the product. Data requirements (set forth at 40 C.F.R. Part 158) vary, but can include product chemistry; mammalian toxicity; environmental toxicity and fate; and residue chemistry, reentry exposure, and spray drift. Efficacy studies generally are not required to be submitted, except for certain antimicrobial pesticides, but must be submitted upon EPA request.

FIFRA Section 3(c)(2)(B) authorizes the EPA to require additional new studies from current registrants "to maintain in effect an existing registration of a pesticide." A "Data Call-In" (DCI) is directed to affected registrants and specifies the additional tests that the EPA requires. Registrants may individually submit, jointly develop, or share in the cost of developing those data.

Under FIFRA Section 3(c)(1)(F)(i), data submitters are given a 10-year period of exclusive use for data submitted in support of a registration for: (1) a new pesticide chemical, or (2) new uses of an already-existing pesticide. The exclusive use provision applies only to data submitted to support an active ingredient first registered after September 30, 1978. A registrant may not rely on exclusive use data without the data owner's consent. The 10-year exclusive use period begins on the date of first registration of the new active ingredient. No exclusive use rights attach to data submitted in response to a DCI. The Food Quality Protection Act (FQPA) extended exclusive use time periods for minor uses, and extended exclusive use protection to data in support of a tolerance or tolerance exemption. These exclusive use protections are particularly relevant to innovators of nanopesticides in that they offer 10-year markets for any active ingredient considered "new."

4.2.2   EPA OPP Nanotechnology Initiatives

The EPA's Office of Pesticide Programs (OPP) is working with other EPA program offices in considering how best to address the growing number of issues that engineered nanoscale materials pose. These OPP initiatives are discussed below.

4.2.2.1   The EPA White Paper

The EPA *Nanotechnology White Paper* includes a discussion of FIFRA. The EPA notes its expectation that "[p]esticide products containing nanomaterials will be subject to FIFRA's review and registration requirements" [24].* The EPA also observes that nanotechnologies may produce "[m]ore-targeted fertilizers and pesticides that result in less agricultural and lawn/garden runoff of nitrogen, phosphorous, and toxic substances is potentially an important emerging application of nanotechnol[o]g[ies] that can contribute to sustainability" [24].

4.2.2.2   OPP Nanotechnology Workgroup

The OPP formed a Nanotechnology Workgroup in late 2006 that is specifically tasked with developing a regulatory framework that will address the nanomaterial pesticide issues that arise under FIFRA. The OPP can be expected to address several core issues in the context of developing its nanotechnology framework.**

A threshold question that the OPP is considering is whether a nanoscale version of a registered conventional pesticide also is considered a registered pesticide. This FIFRA question is similar to the question under the TSCA as to whether a nanoscale version of an existing TSCA Inventory-listed chemical substance also is considered an existing chemical substance. Because of basic differences in the statutory design of FIFRA and TSCA, however, the answer under FIFRA is considerably clearer. As noted above, under FIFRA Section 3(c)(5)(D), registration decisions depend on an EPA determination that a pesticide "will not generally cause unreasonable adverse effects on the environment." In making this determination with respect to nanoscale substances, the EPA must assess whether the benefits of a nanopesticide outweigh its risks, and must determine the conditions under which a nanopesticide may be registered to limit any risks appropriately. Factors in that determination include the composition of the nanopesticide, and claims made with regard to its application and efficacy. Because the balancing of risks and benefits of a nanopesticide is likely different from that for a corresponding registered conventional pesticide, it is probable that the EPA would take the position that use of a nanoscale ingredient in place of its conventional counterpart in a registered pesticide would require the need to submit a new or amended registration. The EPA has taken no official position on this issue, however.

The heart of the EPA's authority under FIFRA to regulate nanopesticides is the registration requirement of FIFRA Section 3. FIFRA prohibits the sale or distribution of unregistered pesticides. As noted, the EPA requires registration applicants to develop extensive information relevant to an assessment of the pesticide's risks and benefits. Thus, through registration requirements, the EPA can prohibit the use of nanopesticides that are determined to present "unreasonable adverse effects" on

---

* Nanotechnology White Paper at 66. In a November 2006 presentation to the Pesticide Program Dialogue Committee, a federal advisory committee that provides advice and recommendations to OPP on pesticide issues, OPP explained that FIFRA's no unreasonable adverse effects finding "must be made regardless of size and whether or not [a product] is engineered or naturally occurring (i.e., all pesticide products are held to the same standard)." OPP, Presentation on Nanotechnology to the Pesticide Program Dialogue Committee (Nov. 9, 2006) at 22, available at http://www.epa.gov/oppfead1/cb/ppdc/2006/november06/session7-nanotec.pdf.

** For a more detailed review of nanotechnology and FIFRA, see ABA, SEER, The Adequacy of FIFRA to Regulate Nanotechnology-Based Pesticides (May 2006), available at http://www.abanet.org/environ/nanotech/pdf/FIFRA.pdf; J. Kuzma and P. VerHage, Nanotechnology in Agriculture and Food Production — Anticipated Applications, Woodrow Wilson International Center for Scholars, Project on Emerging Nanotechnologies (September 2006)

human health or the environment, and may restrict other nanopesticides to ensure that any potential risks do not become unreasonable consistent with EPA's authority under FIFRA Section 6(a)(2).

The inclusion of nanoscale materials as inert ingredients in pesticide formulations also raises interesting and, to date, unanswered questions. It is not clear what the review process will be for a new inert ingredient and/or the nanoscale version of an existing inert ingredient, what data requirements might apply, and what process the OPP will use to review these registered issues. The OPP's Nanotechnology Workgroup is expected to shed light on these issues.

4.2.2.3 Nanotechnology and Antimicrobials

In a late 2006 regulatory status update that was widely reported in the trade press, OPP announced that it had informed Samsung Electronics that a silver ion generating washing machine, which the company had been marketing with claims that it would kill bacteria on clothing, is subject to registration as a pesticide under FIFRA.* The OPP indicated then that a forthcoming *Federal Register* notice "will outline and clarify the Agency's position on the classification of machines that generate ions of silver or other substances for express pesticidal purposes," and that the notice will "not represent an action to regulate nanotechnology" because the EPA "ha[s] not yet received any information that suggests [the Samsung washing machine] involves the use of nanomaterial."** Should the OPP receive such information in the context of a FIFRA registration application, it is expected that the OPP would review the application with the same degree of scrutiny and scientific rigor that it would apply to any other registration application submitted under FIFRA Section 3(c)(5), which establishes the criteria for a pesticide's registration.

The EPA issued its clarifying notice on September 21, 2007 [28]. In the notice, the EPA clarifies that the key distinction between pesticides and devices is whether the pesticidal activity of the article is due to physical or mechanical actions, or due to a substance or mixture of substances. The EPA states that ion generating machines that incorporate a substance, such as silver or copper, in the form of an electrode, and that pass a current through the electrode to release ions of that substance for the purpose of preventing, destroying, repelling, or mitigating a pest are considered by the EPA to be pesticides for FIFRA purposes, and must be registered prior to sale or distribution. The EPA's notices set forth a detailed timeline for affected entities to obtain appropriate EPA approvals and revised labeling, which should be reviewed carefully to avoid enforcement consequences.

Despite press reports to the contrary, the ion generating debate is less about nanopesticides than it is about the EPA's evolving thinking on what constitutes a "device" for FIFRA purposes and thus need not be registered as a pesticide product. The OPP is, however, plainly focusing on nanopesticides and how best to assert the

---

* See OPP, "Regulatory Status Update: Ion Generating Washing Machines" (December 6, 2006), available at http://www.epa.gov/oppad001/ion.htm. Shortly after OPP issued its announcement, the Natural Resources Defense Council (NRDC) wrote to the OPP Director and applauded the "recent decision to regulate the use of nanosilver as a pesticide under [FIFRA]." NRDC Letter to Jim Jones, OPP (November 22, 2006), Available at http://www.nrdc.org/media/docs/061127.pdf.

** See previous footnote and accompanying text.

EPA's jurisdiction over nanopesticides under FIFRA. For example, as of this writing, the EPA is expected to revise the pesticide registration application to require pesticide particle size information, a data field that heretofore the EPA has not required to be completed. It is not clear if this information will be sought with respect to active ingredients only, or active ingredients and any inert ingredient included in a pesticide formulation.

New agricultural/antimicrobial products and application techniques are likely to revolutionize these markets, and there are many commercial opportunities to promote sustainable agricultural and pollution prevention through nanotechnologies. Industry stakeholders and others must engage with the EPA and the U.S. Department of Agriculture early, openly, and regularly to ensure nanotechnologies fulfill their promise as pollution prevention and sustainable agricultural tools.

4.3   THE CLEAN AIR ACT (CAA)

4.3.1   CAA STATUTORY AND REGULATORY BACKGROUND

The Clean Air Act (CAA) is an important statute for controlling environmental impacts of nanotechnology given the potential implications for human health of airborne nanoparticles. Due to their size, ambient nanoparticles may be especially effective in producing respiratory inflammation. The discussion below identifies the likeliest CAA pathway that the EPA and other regulatory agencies might use, as well as their respective limitations as workable regulatory tools for managing emissions from applied nanotechnology.* The EPA's Office of Air and Radiation (OAR) has issued little to no information regarding how it intends to approach regulating nanoscale materials. Several statutory provisions would appear to provide the EPA with the authority to regulate nanoscale substances and the CAA. Each is discussed below.

4.3.1.1   National Air Quality Standards for Particulates Under CAA Sections 108 and 109

CAA Section 109 requires the EPA to establish national ambient air quality standards (NAAQS) for each of the so-called "criteria" pollutants identified by the EPA in Section 108. These two provisions were the drivers that helped power the CAA in the early years after its 1970 enactment. Section 108(a)(1) directs the EPA to publish, and periodically to revise, a list of air pollutants from "numerous or diverse mobile or stationary sources," the emissions of which "cause or contribute to air pollution

---

* For a more detailed review of CAA and nanotechnology, see ABA, SEER, CAA Nanotechnology Briefing Paper (June 2006), available at http://www.abanet.org/environ/nanotech/pdf/CAA.pdf.

which may reasonably be anticipated to endanger public health or welfare."* Section 108(a)(2) directs the EPA to publish air quality "criteria" for each listed pollutant that will "accurately reflect the latest scientific knowledge useful in indicating the kind and extent of all identifiable effects on public health or welfare which may be expected from the presence of such pollutant in the ambient air" [29]. The commonly used term "criteria pollutant" derives from this provision.

Section 109 requires the EPA, based on the air quality criteria in Section 108, to promulgate numerical "primary" and "secondary" NAAQS for each such criteria pollutant. Under Section 109(b)(1), a primary standard is one that will protect the public health, "allowing an adequate margin of safety" [30]. A secondary standard is one that is intended to protect the public welfare.** It is settled law that considerations of cost or technological feasibility are not to play a role when the EPA establishes NAAQS for a pollutant [31–33].

NAAQS have been established for six criteria pollutants — ozone, particulate matter (PM), sulfur dioxide, nitrogen oxide, carbon monoxide, and lead. Among these, it is the PM standards that offer a possible pathway for regulating nanoparticle emissions under the CAA. Observing that particles as a class "span many sizes and shapes and consist of hundreds of different chemicals," the EPA describes PM as "a highly complex mixture of solid particles and liquid droplets distributed among numerous atmospheric gases that interact with solid and liquid phases" [34]. Ambient nanoparticles are the smallest among them.

The EPA's original NAAQS for PM did not make distinctions by particle size, but covered all PM under one primary standard and one secondary standard established for "total suspended particulate" (TSP). Subsequently, as scientists and regulators focused their attention on the potential health effects, and also the impacts on visibility, associated with finer — as opposed to coarser — particles in the air, the EPA made fundamental changes in the PM standards. In 1987, the EPA adopted a final rule that replaced the TSP measure with standards written in terms of PM_{10}, that is, particles with a diameter no greater than 10 micrometers (μm).

Ten years later, the EPA restructured the NAAQS for PM. The EPA's 1997 revision divided the PM universe by size for standard-setting purposes into two groups: (1) "inhalable coarse particles" ($PM_{10-2.5}$), those between 2.5 and 10 μm in diameter; and (2) fine particles ($PM_{2.5}$), those with a diameter of 2.5 μm or smaller. For $PM_{2.5}$, the EPA established primary NAAQS of 15 μg/m³ (annual standard) and 65 μg/m³

---

* 42 U.S.C. § 7408(a)(1)(A) and (B). The term "air pollutant" is defined broadly in Section 302(g), 42 U.S.C. § 7602(g), to mean "any air pollution agent or combination of such agents, including any physical, chemical, biological, radioactive (including source material, special nuclear materials, and by-product material) substance or matter which is emitted into or otherwise enters the ambient air. Such term includes any precursors to the formation of any air pollutant, to the extent the Administrator has identified such precursor or precursors for the particular purpose for which the term 'air pollutant' is used."

** CAA § 109(b)(2), 42 U.S.C. § 7409(b)(2). No "margin of safety" is called for in establishing a secondary NAAQS.

(24-hour average). Challenged by various industry petitioners, the $PM_{2.5}$ standards eventually were upheld in 2002.*

On January 17, 2006, the EPA proposed new revisions to the $PM_{2.5}$ standards, under a schedule that called for issuing final standards by September 27, 2006.** If the primary standards are adopted as proposed, the 24-hour standard will be tightened from 65 to 35 $\mu g/m^3$, while the 15-$\mu g/m^3$ annual standard will be retained. The secondary standards as proposed would be the same as the primary standards.

Neither the December 20, 2005, proposed revised PM standards, nor the background documents issued together with the proposal, discuss the standards in the context of particles emitted from applied nanotechnology. This omission may reflect little more than that nanotechnology and its implications for federal regulators may have been scarcely a blip on the radar screen when development of the revised NAAQS began.*** It does, however, at least indicate that a regulatory strategy to address airborne emissions from applied nanotechnology was not front and center among the EPA's goals in drafting the revised $PM_{2.5}$ standards.

4.3.1.2   Hazardous Air Pollutant Standards Under CAA Section 112

The standards for regulating hazardous air pollutants (HAPs) issued by the EPA under CAA Section 112 offer another pathway for regulating emissions from industries involved in nanotechnology [35]. In contrast to Section 108, Section 112 does not contain a threshold requiring "numerous and diverse" sources to trigger federal regulation. Thus, it is available to address pollutants that are not necessarily ubiquitous nationwide. Section 112 allows the EPA to target pollutants of concern on an industry-wide basis, from both new and existing stationary sources, once they are listed as HAPs under Section 112(b). Congress identified an initial list of 189 pollutants into the law. The EPA is authorized to add pollutants to the list (or to remove them) on its own initiative or in response to a third-party petition.

Congress set a 10-year deadline of November 2000 for the EPA to adopt the required technology-based emission standards for the universe of major industrial

---

* The 1997 PM standards, together with controversial revisions to the ozone NAAQS promulgated at the same time, were the subject of protracted litigation in the Court of Appeals for the District of Columbia Circuit, the Supreme Court, and, finally, again in the D.C. Circuit, which ultimately upheld them. See American Trucking Assn v. EPA, 175 F.3d 1027 (D.C. Cir. 1999); Whitman v. American Trucking Ass'n, 531 U.S. 457 (2001); and American Trucking Ass'n. v. EPA, 283 F.3d 355 (D.C. Cir. 2001), respectively.

** 71 Fed. Reg. 2620 (Jan. 17, 2006). The schedule for completion of this review is the result of a lawsuit initiated by the American Lung Association and other plaintiffs in 2003 to enforce the 5-year cycle established in CAA Section 108(d) for EPA to review the NAAQS and make any needed revisions. See American Lung Ass'n. v. Whitman, No. 03-778 – ESH (D.D.C.).

*** In its proposal, the EPA seeks comment on a variety of alternatives to various aspects of the proposal. Conceivably, it could decide to specifically target the smallest among the universe of PM2.5 particles. The preamble to the proposal states, however, that "the Administrator provisionally concludes that currently available studies do not provide a sufficient basis for supplementing mass-based fine particle standards for any specific fine particle component or subset of fine particles, or for eliminating any individual component or subset of components from fine particle mass standards." 71 Fed. Reg. at 2645.

source categories, as well as for area sources.* These maximum achievable control technology (MACT) standards incorporate "floor" requirements and are defined to require the "maximum degree of reductions and emissions deemed achievable for the [industrial source] category or subcategory" that the EPA, "taking into consideration the cost of achieving the reduction, any non-air-quality health and environmental impacts and energy requirements, determines is achievable for new or existing sources" [38].

These control technologies may include process or material changes; enclosures; collection and treatment systems; design, equipment, work practice, or operational changes; or a combination of the foregoing [38]. For area sources, the EPA has the option to establish alternative standards that do not necessarily rise to the stringency of what MACT requires. For these sources, Section 112(d)(5) provides for "the use of general available control technologies [GACT] or management practices" [39], which does not necessitate setting a minimum control level that might prove daunting for non-major sources to meet in practice. Although the EPA did not meet the 10-year deadline for promulgation of MACT standards for all current subject source categories, most by now are in effect, and the EPA has covered a great deal of regulatory ground in the process.

Section 112(f) provides a second, health-based line of defense for MACT sources, in the form of "residual risk" emissions standards. These are to be established within 8 years after MACT standards are promulgated for a source category, if the EPA determines, following a risk assessment, that such standards are necessary. Where they apply, residual risk standards, similar to the pre-1990 HAP standards, must incorporate an "ample margin of safety to protect public health" [40]. Because the task of promulgating MACT standards went beyond the November 2000 deadline, residual risk standard-setting still is in its early stages, and it is too soon to determine the real-world impact, including compliance issues, that these health-based standards will have.

In its cursory summary of Section 112, the EPA *Nanotechnology White Paper* notes, but does not elaborate on, the provisions of Section 112(r) that are intended to prevent the accidental release of extremely hazardous substances and to minimize the consequence of any such release that should occur [41]. An "accidental release" is defined as "an unanticipated emission of a regulated substance or other extremely hazardous substance into the ambient air from a stationary source" [42]. The EPA was directed to establish an initial list of the 100 substances posing the greatest risk of causing death, injury, or serious adverse effects to human health or the environment in the event of such an accidental release, along with threshold quantities that, if released, would set the Section 112(r) provisions in motion [43]. The White Paper, however, does not elaborate on whether the EPA views the accidental release provision as particularly significant in the context of regulating

---

* CAA § 112(e)(1), 42 U.S.C. § 7412(e)(1). A "major source" is defined as "any stationary source or group of stationary sources located within a contiguous area and under common control that emits or has the potential to emit considering controls" 10 tons per year (TPY) of any single HAP or 25 TPY of any combination of HAPs. CAA § 112(a)(1), 42 U.S.C. § 7412(a)(1). An "area source" is any non-major stationary source of HAPs; it expressly excludes motor vehicles. CAA § 112(a)(2), 42 U.S.C. § 7412(a)(2).

nanotechnology. Any future addition of nanoparticles to the Section 112(r) list would need to be based on risk assessment data that go well beyond what are currently available. Presumably, any such listing would be accompanied by the EPA's establishing a very small threshold release quantity, commensurate with the nanomaterials at issue. It is unlikely that the EPA has given substantial thought at this juncture to the role that Section 112(r) might play in this context.

Section 112 may offer a better fit for the future regulation of nanoparticle emissions than do the particulate NAAQS established under Sections 108 and 109, although questions necessarily remain. For the current universe of MACT sources, since Congress provided an initial list of nearly 200 pollutants, the EPA was able to skip over the HAP identification and listing issue that triggers regulation in the first place. Unless nanotechnology-associated production processes generate pollutants already listed under Section 112, the EPA would have to determine whether — and which — nanoparticles meet the test for listing. The process of adding a pollutant to the Section 112 list, which is accomplished through rulemaking, must be based on a body of data that, at this point, is unlikely to exist. Accordingly, listing, in the nanotechnology context, realistically must await a more robust database.

4.3.1.3   Fuel Additives under CAA Section 211

CAA Section 211 requires all fuels and fuel additives distributed in commerce in the United States to be registered by the EPA. In the past, obtaining and maintaining an EPA registration for a fuel or fuel additive was often a relatively simple process. This process, however, has become more complex in recent years, as the EPA has introduced requirements for complex testing to support fuel and fuel additive registrations. The EPA also has increased its scrutiny of the impact of fuel and fuel additive products on public health and welfare, and on the increasingly elaborate devices and systems it requires to control motor vehicle emissions, in no small part because certain more recent fuel additives have contained nanoscale metal substances. To the extent these nanoscale metals have proven efficiency as fuel additives, the EPA can be expected to use CAA Section 211 to authorize obtaining additional testing. To date, however, the EPA has not disclosed publicly what exactly it is up to in this regard. The EPA *Nanotechnology White Paper* notes [24]:

> EPA's Office of Air and Radiation/Office of Transportation and Air Quality has received and is reviewing an application for registration of a diesel additive containing cerium oxide. Cerium oxide nanoparticles are being marketed in Europe as on- and off-road diesel fuel additives to decrease emissions and some manufacturers are claiming fuel economy benefits.

4.4   THE CLEAN WATER ACT (CWA)

4.4.1   CWA Statutory and Regulatory Background

Like the CAA, the Federal Water Pollution Control Act, more commonly known as the Clean Water Act (CWA), is an important media-specific statute for controlling the environmental impacts of nanoscale substances. The CWA governs discharges of "pollutants" into waterbodies, more particularly into "waters of the United States."* As in the CAA, the statutory definition of a "pollutant" is expansive,** and likely includes engineered nanoscale materials and engineered nanoscale material-containing wastewaters. The stated objective of the CWA is "to restore and maintain the chemical, physical, and biological integrity of the Nation's waters" [44].

In its *Nanotechnology White Paper*, the EPA states that "[d]epending on the toxicity of nanomaterials to aquatic life, aquatic dependent wildlife, and human health, as well as the potential for exposure, nanomaterials may be regulated under the CWA" [24]. The EPA points out that "[a] variety of approaches are available under the CWA to provide protection, including effluent limitation guidelines, water quality standards ...., best management practices, [point source discharge] permits, and whole effluent toxicity testing" [24]. Below is a discussion of the more prominent of these approaches.***

4.4.2 THE NATIONAL POLLUTANT DISCHARGE ELIMINATION SYSTEM (NPDES) PROGRAM

The centerpiece of the CWA regulatory program is the National Pollutant Discharge Elimination System (NPDES) established under Section 402 of the statute. The key features of the NPDES program are

- The issuance, by either the EPA or a state with an EPA-approved permitting program, of point source discharge permits containing numeric, pollutant-specific effluent limitations that either are technology-based or water quality-based****
- Routine and frequent monitoring of effluent (i.e., wastewater) through sampling and analytical methods to determine compliance
- Routine and frequent reporting to the permitting authority of the permittee's effluent monitoring results

---

* The CWA actually covers discharges into "navigable waters," which are defined as "waters of the United States, including the territorial seas." CWA § 502(7), 33 U.S.C. § 1362(7). In its CWA implementing regulations, the EPA defines the phrase "waters of the United States" in an extremely broad fashion. See EPA 40 C.F.R. § 122.2.

** Section 502(6) of the CWA defines the term "pollutant" to mean "dredged spoil, solid waste, incinerator residue, sewage, garbage, sewage sludge, munitions, chemical wastes, biological materials, radioactive materials, heat, wrecked or discarded equipment, rock, sand, cellar dirt and industrial, municipal, and agricultural waste discharged into water." 33 U.S.C. § 1362(6). A "discharge of a pollutant" is defined in relevant part as "any addition of any pollutant to [waters of the United States] from any point source," with the term "point source" defined broadly to mean "any discernible, confined and discrete conveyance, including but not limited to any pipe, ditch, channel, tunnel, conduit, well, discrete fissure, container, rolling stock, concentrated animal feeding operation, or vessel or other floating craft, from which pollutants are or may be discharged." CWA §§ 502(12), 502(14), 33 U.S.C. §§ 1362(12), 1362(14).

*** For a more detailed review of nanotechnology and the CWA, see ABA, SEER, Nanotechnology Briefing Paper: Clean Water Act (June 2006), available at http://www.abanet.org/environ/nanotech/pdf/cwa.pdf.

**** Technology-based effluent limitations derive from CWA Sections 301 and 304, while water quality-based effluent controls stem from Section 302.

Under CWA Section 301(a), it is unlawful for a person to discharge any pollutant into the waters of the United States "except as in compliance with" an NPDES permit [45].

Wastewater containing nanoscale materials is subject to effluent limitations, whether technology-based or water quality-based, set forth in an NPDES permit. To date, however, the EPA has not released publicly how it intends to develop effluent limitations specifically for engineered nanoscale material-containing wastewaters, or even if it intends to do so. Nor has it given any indication as to whether engineered nanoscale materials constitute conventional, nonconventional, or toxic pollutants, a distinction that bears directly on the type of technology that a permitted discharger must employ to achieve a particular effluent limitation. Little currently is known about the availability and economic feasibility of technology to control wastewater discharges containing engineered nanoscale materials.

4.4.3 PRETREATMENT STANDARDS

The NPDES permit program applies to so-called direct dischargers — that is, facilities that discharge pollutants directly to waters of the United States. It does not apply to what are known as indirect dischargers — that is, facilities that discharge wastewater to publicly owned treatment works (POTWs) rather than directly to waterbodies [46]. The EPA's pretreatment program, mandated by CWA Section 307(b), establishes pretreatment standards for this latter category of dischargers [47].

As with effluent limitations, it would appear that the EPA is considering these issues but has yet to release any information on its development and issuance of pretreatment standards specific to nanoscale material-containing wastewater streams. It bears noting, however, that the OPP's December 2006 determination, discussed above,* that Samsung Electronics' silver ion generating washing machine warrants registration as a pesticide under FIFRA was precipitated in large part by letters sent to the OPP by the National Association of Clean Water Agencies (NACWA) and an organization representing California POTWs. The NACWA and the POTWs were concerned about the discharge of silver ions to wastewater treatment plants.**

4.5   THE RESOURCE CONSERVATION AND RECOVERY ACT (RCRA)

4.5.1   RCRA STATUTORY AND REGULATORY BACKGROUND

The Resource Conservation and Recovery Act (RCRA) manages the generation, transport, and disposal and recycling of materials defined as "hazardous waste." The EPA is well aware of the potential promise found in nanotechnology applications to detect, monitor, and clean up environmental contaminants. Many of the EPA's resources to date have been devoted to this aspect of nanotechnology, as opposed

---

*See OPP, "Regulatory Status Update: Ion Generating Washing Machines" (December 6, 2006), available at http://www.epa.gov/oppad001/ion.htm. Shortly after the OPP issued its announcement, the Natural Resources Defense Council (NRDC) wrote to the OPP Director and applauded the "recent decision to regulate the use of nanosilver as a pesticide under [FIFRA]." NRDC Letter to Jim Jones, OPP (November 22, 2006), Available at http://www.nrdc.org/media/docs/061127.pdf.

** See, e.g., Letter to Jim Jones, OPP, from Chuck Weir, Tri-TAC (January 27, 2006), available at http://www.tritac.org/documents/letters/2006_01_27_EPA_Samsung_Silver_ Wash.pdf.

to addressing how RCRA and the EPA's implementing regulations might apply to nanowaste. The RCRA implications of nanotechnology are less well defined.*

In determining under RCRA whether a material is a hazardous waste, the first step is to determine whether it is a solid waste. Section 1004(5) of RCRA defines the term "hazardous waste" to mean:

> A solid waste, or combination of solid wastes, which because of its quantity, concentration, or physical, chemical, or infectious characteristics may (a) cause, or significantly contribute to an increase in mortality or an increase in serious irreversible, or incapacitating reversible, illness; or (b) pose a substantial present or potential hazard to human health or the environment when improperly treated, stored, transported, or disposed of, or otherwise managed.

If a waste is considered a solid waste, the next step is to determine whether the waste is specifically excluded from RCRA regulation, is a "listed" hazardous waste, or is a "characteristic" waste. Congress excluded several classes of solid waste from regulation as hazardous waste. Congress did not exclude these materials because they are inherently different or "less hazardous" than other materials deemed solid waste, but because it lacked data enabling it to determine whether these materials should be regulated as hazardous. In some cases, the EPA excluded a solid waste from regulation as hazardous waste after determining it would be impractical, unfair, or otherwise undesirable to regulate the waste as hazardous. As discussed below, an important category of waste currently excluded from registration is household hazardous waste. Given the increasing number of consumer products enabled by nanotechnology, this class of exempt waste arguably could exclude larger quantities of nanoscale waste from being regulated under RCRA when disposed.

4.5.2 LISTED HAZARDOUS WASTES

Under RCRA Subtitle C, a solid waste can be a hazardous waste in one of two ways. First, a solid waste that is "listed" — that is, appears on one of three lists found in 40 C.F.R. Part 261, Subpart D — is a hazardous waste unless excluded as the result of a petition to delist filed by an interested party. Second, a solid waste that is not listed can still be considered hazardous if it exhibits one of four characteristics: ignitability, corrosiveness, reactivity, or toxicity.

Currently, there are three lists of hazardous waste:

1. Hazardous Wastes from Nonspecific Sources (found at Section 261.31 and commonly called the "F list")
2. Hazardous Wastes from Specific Sources (at Section 261.32 and known as the "K list")

---

* For a more detailed review of these issues, see ABA, SEER, RCRA Regulation of Wastes from the Production, Use, and Disposal of Nanomaterials (June 2006), available at http://www.abanet.org/environ/nanotech/pdf/RCRA.pdf; see also L. K. Breggin and J. Pendergrass, Where Does The Nano Go? End-of-Life Regulation of Nanotechnologies, Woodrow Wilson International Center for Scholars, Project on Emerging Nanotechnologies (July 2007).

3. Unused Discarded Commercial Chemical Products (specified at Section 261.33, and including acutely hazardous wastes in the "P list" and non-acutely hazardous wastes in the "U list")

Each listing is accompanied by a background document, which describes EPA's basis for listing the waste. These background documents often are helpful in determining the applicability of a listing. A key issue that waste generators must confront is whether a waste listing includes a nanoscale version that may be fundamentally different from its conventionally sized counterpart. The EPA has not yet issued guidance on this issue.

None of the EPA's hazardous waste listings explicitly includes or discusses nanoscale materials. Certain waste listings likely will include nanoscale materials or wastes from certain nanomaterials manufacturing. For example, K-listed wastes include discarded materials from organic chemicals, pesticides, and inorganic chemicals. As noted above under TSCA and FIFRA, nanoscale chemicals and pesticides could well be generated by these industries, and thus the K-listed codes presumably could include these nanomaterials. Again, the EPA has issued no guidance on these issues.

4.5.3 Characteristic Hazardous Waste

Solid wastes that are not listed can still be considered hazardous if they exhibit one or more of the hazardous waste characteristics: ignitability, corrosivity, reactivity, or toxicity. The first three characteristics refer to properties of the waste itself, while the fourth evaluates a waste's potential to release certain hazardous constituents when disposed.

The EPA intended the Toxicity Characteristic (TC) to reflect the potential for leaching to groundwater that results from the co-disposal of toxic wastes in an actively decomposing municipal landfill generating an acidic leachate. The EPA requires the application of an extraction test — the Toxicity Characteristic Leaching Procedure (TCLP) — to determine if a waste leaches any of the 39 specified toxicants above regulatory thresholds. Any leachate sample created using the TCLP that contains a regulated constituent in concentrations at or exceeding its regulatory threshold exhibiting the toxicity characteristic is considered by the EPA to be a hazardous waste.

While the EPA's Office of Solid Waste and Emergency Response (OSWER) has yet to issue any regulatory pronouncements with regard to nanoscale waste, there is no reason to believe these materials would be treated differently than any other waste materials for purpose of RCRA waste classification. It is likely that many nanoscale materials will display one or more of these characteristics, due to their inherent composition, and thus will likely constitute characteristically hazardous waste under RCRA upon disposal. An often-cited example is nanoscale aluminum, which is combustible. If disposed without treatment to eliminate its combustability, this material would likely qualify as a characteristically ignitable hazardous waste.*

---

* EPA 40 C.F.R. § 261.21(a)(2) (non-liquid wastes are characteristically ignitable if they "are capable, under standard temperature and pressure, of causing fire through friction, absorption of moisture or spontaneous chemical changes and, when ignited, burns so vigorously and persistently that it creates a hazard").

Other nanoscale materials would appear to include toxic constituents that, upon disposal, would likely render the materials characteristically toxic and thus subject to RCRA upon disposal.*

As noted, OSWER has not yet provided any guidance on these issues. Similarly, OSWER has not commented on whether the application of conventional RCRA testing procedures as set forth in *Test Methods for Evaluating Solid Waste, Physical/Chemical Methods (SW-846)* are well suited to nanoscale materials. It is believed, however, that some of the unique qualities of nanoscale materials may materially affect the results of RCRA waste testing procedures.

4.5.4 MIXTURE AND DERIVED-FROM RULES

Under RCRA's rule, if a listed hazardous waste is mixed with other material, the entire mixture assumes the status of the listed waste.** While RCRA requirements do not explicitly apply to nanoscale materials, there is no reason to believe that they do not. Accordingly, if a listed hazardous waste is processed in a way that causes it to generate a sludge, spill residue, ash emission control dust, leachate, or other form of solid waste [48], then under the "derived-from rule," the resulting solid waste assumes the same listed waste code as the original listed hazardous waste.

If a characteristic or listed hazardous waste is spilled into soil or another environmental medium, under the "contained-in" principle, the resulting mixture of soil and hazardous waste is deemed to "contain" the hazardous waste until it has been treated to a point where the soil no longer contains the hazardous waste.*** The EPA has not issued any guidance on whether, under the contained-in principle, the presence of a nanoscale material qualified as a listed hazardous waste would trigger application of the contained-in rule. There is no reason to believe that nanoscale versions of listed hazardous wastes, mixtures of large amounts of solid wastes may become listed hazardous waste because they contain small amounts of nanoscale listed hazardous waste.

The EPA included several exemptions from the definitions of both "solid waste" and "hazardous waste." These exclusions include, among others, household waste, certain fertilizers made from hazardous wastes, and other materials listed in 40 C.F.R. § 261.4(a), as well as certain agricultural wastes returned to the soil as fertilizer. As

---

* Quantum nanodots typically often contain cadmium or selenium. The EPA has designated wastes yielding more than 1.0 milligrams per liter of cadmium or selenium through a TCLP extraction test as characteristically toxic upon disposal (i.e., D006 or D010 waste). EPA 40 C.F.R. § 261.24(b).

** EPA 40 C.F.R. § 261.3(a)(2)(iv). The mixture rule includes exemptions for (i) mixtures that include hazardous wastes listed solely as ignitable, corrosive, or reactive; and (ii) *de minimis* amounts of listed hazardous wastes mixed in permitted wastewater treatment systems. The mixture rule also applies to mixtures of characteristic hazardous wastes, but only if the resulting mixture still displays the original hazardous characteristic. EPA 40 C.F.R. § 261.3(a)(2)(i).

*** While the EPA has not promulgated the contained-in principle as a formal regulation for contaminated media, it has issued several guidances to outline its policy. See, e.g., 63 Fed. Reg. 28622 (May 26, 1998); 61 Fed. Reg. 18795 (Apr. 29, 1996). The U.S. Circuit Court of Appeals for the D.C. Circuit has upheld the EPA's application of the contained-in policy. Chemical Waste Management v. EPA, 869 F.2d 1526 (D.C. Cir. 1989). The EPA has also codified the contained-in principle in its rules for debris management.

nanoscale materials become more prevalent in consumer products, one can expect increasing questions regarding the prudence of these exemptions [49].

4.5.5 TRANSPORTER REQUIREMENTS

The hazardous waste transporter regulations at 40 C.F.R. Part 263 are designed to ensure the safe transport of hazardous wastes from generators to treatment, storage, and disposal facilities, or other appropriate destinations. A hazardous waste transporter is defined as "... a person engaged in the offsite transportation of hazardous waste by air, rail, highway, or water" [50]. As stated in this definition, the EPA does not regulate the on-site movement of wastes within a facility's boundaries. Transporters of hazardous wastes generally are subject to regulation under RCRA Subtitle C if the shipment requires a manifest under 40 C.F.R. Part 262.

RCRA hazardous wastes are considered "hazardous materials" under the U.S. Department of Transportation (DOT) regulations. Therefore, hazardous waste transporters also are subject to DOT regulations [51]. RCRA regulations are intended to be consistent with the DOT requirements.

The RCRA Subtitle C requirements for transporters include:

- Obtaining an EPA identification number
- Complying with the manifest requirements
- Taking appropriate action (including cleanup and reporting) in the event of an accident and/or release of a hazardous waste
- Complying with recordkeeping requirements

In some circumstances, the transporter may wish to store shipments of waste for short periods of time incidental to transport. Transporters are not subject to the 40 C.F.R. Part 264 requirements for hazardous waste storage facilities if a manifested shipment of hazardous waste is stored at a transfer facility for 10 days or less in containers that comply with 40 C.F.R. Section 262.30. If hazardous wastes are stored at a transfer facility for more than 10 days, however, the transfer facility becomes a hazardous waste storage facility subject to Subtitle C permitting and storage facility standards.

Hazardous waste transporters also must comply with all other applicable RCRA Subtitle C requirements. For example, transporters must comply with 40 C.F.R. Part 266 when managing certain recyclable materials or military munitions, and 40 C.F.R. Part 268 when the shipment consists of wastes subject to the land disposal restrictions (LDRs). In addition, transporters of universal wastes must comply with the 40 C.F.R. Part 273 standards. The regulatory requirements for universal waste transporters are contained in Subpart D to Part 273.

The RCRA regulations require that hazardous wastes be shipped in accordance with DOT's hazardous materials regulations (HMRs), at 40 C.F.R. Parts 100 through 185. The DOT regulations apply to parties involved with shipping hazardous materials by highway, rail, air, and water. All shipments of hazardous wastes that are subject to the RCRA manifest requirements are subject to the HMRs. The DOT requirements contain provisions for classifying, packaging, marking, labeling, placarding, and handling hazardous waste shipments. The DOT regulations require that hazardous waste transporters: (1) register; (2) provide employee training; (3) comply with shipping paper/manifest requirements; (4) follow certain procedures before, during, and after the transport of waste; and (5) respond to releases of hazardous waste.

There is no evidence to suggest that large quantities of nanowastes are being transported for disposal. Nonetheless, for the reasons described above, if these materials were being transported and qualified as RCRA hazardous waste, there is no reason to believe that the RCRA transporter requirements would not apply.

4.5.6 TREATMENT, STORAGE, AND DISPOSAL FACILITY REQUIREMENTS

The treatment, storage, and disposal (TSD) requirements for RCRA permitted facilities are set forth in 40 C.F.R. Part 264, and 40 C.F.R. Parts 264 and 265 provide both general requirements for TSD facilities and standards that apply to specific types of TSD waste management units.

The hazardous waste TSD facility standards apply to facilities that treat, store, or dispose of hazardous wastes. As a result, applicability of TSD facility requirements will hinge on definitions of the terms "facility," "treatment," "storage," disposal," and "disposal facility," each of which is defined in 40 C.F.R. Section 260.10.

A RCRA permit, which the EPA or authorized states will issue, gives owners/operators of TSD facilities legal authority to treat, store, or dispose of hazardous waste. A permit specifies the technical and administrative standards with which a facility must comply to manage hazardous waste legally. The permit's standards are based on the types of hazardous waste management units at the facility and the specific waste streams that will be managed at the facility. 40 C.F.R. Part 270 establishes the requirements for obtaining RCRA permits. Facilities constructed after the RCRA Subtitle C regulations were promulgated must apply for and receive a RCRA Part B permit before beginning operations, and thus the facility should be designed and operated to meet the requirements of the full RCRA Subtitle C hazardous waste program.

General TSD standards include general facility standards (Subpart B); preparedness and prevention requirements (Subpart C); contingency plan and emergency procedures (Subpart D); and manifest system, recordkeeping, and reporting requirements (Subpart E). Specific TSD standards apply to specific types of treatment, storage, and disposal activities (e.g., tanks, landfills, surface impoundments); to specific types of equipment (e.g., drip pads, process vents); or to specific wastes (e.g., hazardous waste explosives). If a hazardous waste is to be land disposed, the TSD facility also must comply with the LDR requirements of 40 C.F.R. Part 268.

As above, there is little evidence that nanowastes are being treated, stored, or disposed at RCRA facilities. Among the many issues, EPA's OSWER is considering including the appropriate methods for treating and disposing nanowaste. Much more data and information are needed before these issues can be addressed comprehensively and the TSD rules modified, if at all, as needed.

4.6   THE POLLUTION PREVENTION ACT (PPA)

The Pollution Prevention Act of 1990 (PPA) was enacted in November 1990 and amended through Public Law 107-377 in December 2002. Congress declared it a national policy to address pollution based on "source reduction." The policy established a hierarchy of measures to protect human health and the environment, where multimedia approaches would be anticipated:

1. Pollution should be prevented or reduced at the source.
2. Pollution that cannot be prevented should be recycled in an environmentally safe manner.
3. Pollution that cannot be prevented or recycled should be treated in an environmentally safe manner.
4. Disposal or other release into the environment should be employed only as a last resort and should be conducted in an environmentally safe manner.

The first tier of the hierarchy is source reduction — the preferred strategy for addressing potential environmental issues. Source reduction is defined in the PPA as:

"Any practice which: (1) reduces the amount of any hazardous substance, pollutant, or contaminant entering any waste stream or otherwise released into the environment (including fugitive emissions) prior to recycling, treatment, or disposal; and (2) reduces the hazards to public health and the environment associated with the release of such substances, pollutants, or contaminants."

The PPA required the EPA to establish an office to carry out the functions of the statute. In 1990, the EPA formally established the OPPT. Within this office were initiated two programs, with two different approaches, to meet the spirit of the new national policy: (1) the Design for the Environment (DfE) Program and (2) the Green Chemistry Program.* Under the DfE Program, the EPA works in partnership with industry sectors to improve the performance of commercial processes while reducing risks to human health and the environment. The Green Chemistry Program promotes research to design chemical products and processes that reduce or eliminate the use and generation of toxic chemical substances. In 1998, the EPA complimented these two programs with the Green Engineering Program, which applies approaches and tools for evaluating and reducing the environmental impacts of processes and products.** The EPA is well aware of the pollution prevention opportunity that nanotechnology offers. As noted, nanoscale materials may result in the reduction and/or elimination of conventional pesticides is being applied. Nanoscale chemicals may diminish the amount and/or toxicity of conventionally sized industrial chemicals. The EPA is excited about the many pollution prevention opportunities occasioned by nanotechnology.***

---

* The OPPT, as discussed previously, also implements the EPA's responsibilities under TSCA.
** See http://www.epa.gov/oppt/greenengineering.
*** For more information on green nanotechnology, see K. F. Schmidt, Green Nanotechnology: It's Easier Than You Think, Woodrow Wilson International Center for Scholars, Project on Emerging Nanotechnologies (April 2007).

To explore these opportunities in greater detail, the EPA convened a conference in September 2007 entitled "Pollution Prevention through Nanotechnology."* The conference provided an opportunity to exchange ideas and information on using nanotechnology to develop new ways to prevent pollution. Representatives from industry, academia, nongovernmental organizations, and government discussed current practices and potential research areas in nanotechnology that incorporate the concept of pollution prevention in three major areas: (1) products: less toxic, less polluting, and wear-resistant; (2) processes: more efficient and waste-reducing; and (3) energy and resource efficiency: processes and products that use less energy and fewer raw materials.

The intent of the conference was to address which nanotechnologies show the greatest promise for preventing pollution, the most promising areas of research on pollution prevention applications of nanotechnologies, and ways to promote and encourage pollution prevention in the development and application of nanotechnology.

4.7   THE FEDERAL FOOD, DRUG, AND COSMETIC ACT (FEDCA)

The U.S. Food and Drug Administration (FDA) regulates a wide range of products under the Federal Food, Drug, and Cosmetic Act (FFDCA), including foods, cosmetics, drugs, devices, and veterinary products, some of which may utilize nanotechnology or contain nanomaterials. The FDA has not established its own formal definition of "nanotechnology," although the FDA participated in the development of the NNI definition of "nanotechnology," as did many other agencies. Using that definition, nanotechnology relevant to the FDA might include research and technology development that both satisfies the NNI definition and relates to a product regulated by the FDA.

To facilitate the regulation of nanotechnology products, the FDA has formed a NanoTechnology Interest Group (NTIG) composed of representatives from all FDA Centers. The NTIG meets quarterly to ensure there is effective communication between the Centers. Most of the Centers also have working groups that establish the network between their different components. There are also a wide range of products involving nanotechnologies which are regulated by other federal agencies.

In 2006, the FDA also formed a Nanotechnology Task Force. The Task Force is tasked with determining regulatory approaches that encourage the responsible development of FDA-regulated products that use nanotechnology.

On July 25, 2007, the Nanotechnology Task Force issued a report that addresses regulatory and scientific issues, and offered recommendations for each. The Task Force recommended that the FDA consider developing specific guidance for manufacturers and researchers, including guidance to clarify what information should be provided to the FDA about products, and when the use of nanoscale materials may change the regulatory status of particular products. In its press release announcing the availability of the Task Force report, the FDA stated that, as with other FDA guidance, "draft guidance documents would be made available for public comment prior to being finalized." The Task Force also recommended that the FDA work

---

* See http://www.epa.gov/opptintr/nano/scope.htm.

to assess data needs to better regulate nanotechnology products; develop in-house expertise; ensure consideration of relevant new information on nanotechnology as it becomes available; and evaluate the adequacy of current testing approaches to assess safety, effectiveness, and quality of nanoscale materials. The Task Force report is available at http://www.fda.gov/nanotechnology/taskforce/report2007.pdf.

Importantly, the Task Force recommended that the FDA continue to pursue regulatory approaches that take into account the potential importance of material size and the evolving state of the science. Because one definition for "nanotechnology," "nanoscale material," or a related term or concept "may offer meaningful guidance in one context, the Task Force recommended that definition may be too narrow or broad to be of use in another." The Task Force thus "does not recommend attempting to adopt formal, fixed definitions for such terms for regulatory purposes at this time."

The Task Force's initial findings and recommendations are divided into two sections: (1) a review of the scientific knowledge of the potential effects of nanoscale materials relevant to the FDA's regulation of products, with an assessment of scientific issues relating to the FDA's regulation of products using nanoscale materials; and (2) an assessment of the FDA's regulatory authorities as they apply to FDA-regulated products using nanoscale materials. Each is described below.

4.7.1   SCIENCE ISSUES

The Task Force's initial recommendations relating to scientific issues focus on improving scientific knowledge of nanotechnology to help ensure the FDA's regulatory effectiveness, particularly with regard to products not subject to pre-market authorization requirements. The Task Force also addresses the need to evaluate whether the tools available to describe and evaluate nanoscale materials are sufficient, and the development of additional tools where necessary.

4.7.1.1   Issue: Understanding Interactions of Nanoscale Materials with Biological Systems

The Task Force recommends strengthening the FDA's promotion of, and participation in, research and other efforts to increase scientific understanding to facilitate assessment of data needs for regulated products, including:

- Promoting efforts, and participating in collaborative efforts, to further understanding of biological interactions of nanoscale materials, including, as appropriate, the development of data to assess the likelihood of long-term health effects from exposure to specific nanoscale materials
- Assessing data on general particle interactions with biological systems and on specific particles of concern to the FDA
- Promoting and participating in collaborative efforts to further understanding of the science of novel properties that might contribute to toxicity, such as surface area or surface charge
- Promoting and participating in collaborative efforts to further understanding of measurement and detection methods for nanoscale materials

- Collecting/collating/interpreting scientific information, including the use of data calls for specific product review categories
- Building in-house expertise
- Building infrastructure to share and leverage knowledge internally and externally, seeking to collect, synthesize, and build upon information from individual studies of nanoscale materials
- Ensuring consistent transfer and application of relevant knowledge through the establishment of an agency-wide regulatory science coordination function for products containing nanoscale materials

4.7.1.2   Issue: Adequacy of Testing Approaches for Assessing Safety and Quality of Products Containing Nanoscale Materials

To be marketed, FDA-regulated products must be safe and, as applicable, effective. FDA-regulated products also must meet all applicable good manufacturing practice and quality requirements. Adequate testing methods are needed regardless of whether or not a product is subject to pre-market authorization. Accordingly, the following recommendations are relevant to all categories of FDA-regulated products. The FDA should:

- Evaluate the adequacy of current testing approaches to assess safety, effectiveness, and quality of products that use nanoscale materials.
- Promote and participate in the development of characterization methods and standards for nanoscale materials.
- Promote and participate in the development of models for the behavior of nanoscale particles, *in vitro* and *in vivo*.

The Task Force recommends encouraging manufacturers to consult with the FDA regarding the appropriateness of testing methodologies for evaluating products using nanoscale materials.

4.7.2   Regulatory Policy Issues

The Task Force concluded that the FDA's authorities are generally comprehensive for products subject to pre-market authorization requirements, such as drugs, biological products, devices, and food and color additives, and that these authorities give the FDA the ability to obtain detailed scientific information needed to review the safety and, as appropriate, effectiveness of products. For products not subject to pre-market authorization requirements, such as dietary supplements, cosmetics, and food ingredients that are generally recognized as safe (GRAS), manufacturers are generally not required to submit data to the FDA prior to marketing, and the FDA's oversight capacity is less comprehensive.

The Task Force made various recommendations to address regulatory challenges that may be presented by products that use nanotechnology, especially regarding products not subject to pre-market authorization requirements, taking into account the evolving state of the science in this area. A number of recommendations deal with requesting data and other information about the effects of nanoscale materials on the safety and, as appropriate, the effectiveness of products. Other recommendations suggest that the FDA provide guidance to manufacturers about when the use of nanoscale ingredients may require submission of additional data, change the product's regulatory status or pathway, or merit taking additional or special steps to address potential safety or product quality issues. The Task Force also recommends seeking public input on the adequacy of the FDA's policies and procedures for products that combine drugs, biological products, and/or devices containing nanoscale materials to serve multiple uses, such as both a diagnostic and a therapeutic intended use. The Task Force also recommends encouraging manufacturers to communicate with it early in the development process for products using nanoscale materials, particularly with regard to such highly integrated combination products.

4.7.2.1   Issue: Ability of the FDA to Identify FDA-Regulated Products Containing Nanoscale Materials

Recommendations for consideration include:

- Issue guidance to sponsors regarding identification of the particle size for products subject to pre-market authorization, including over-the-counter (OTC) drugs, and food and color additives; and products not subject to pre-market authorization but for which the sponsor is required to provide notice, or may choose to provide notice.
- When warranted, issue a call for data to identify: OTC drug products that contain or may contain nanoscale versions of ingredients included in an OTC monograph; and nanoscale versions of previously approved food and color additives.

4.7.2.2   Issue: Scope of the FDA's Authority Regarding Evaluation of Safety and Effectiveness

For products subject to pre-market authorization:

- Issue a notice in the *Federal Register* requesting submission of data and other information addressing the effects on product safety and effectiveness of nanoscale materials in products subject to FDA pre-market authorization.
- Issue guidance requesting submission of information on whether and how the presence of nanoscale materials affects the manufacturing process for products subject to pre-market authorization, as part of a pre-market submission.
- Issue guidance or amend existing guidance to describe what additional or distinct information should be submitted to the FDA or generated with regard to the following:
  - New food or color additives made with nanoscale materials
  - Previously approved food or color additives that are now made with nanoscale materials or contain greater proportions of nanoscale materials
- Issue guidance describing when:

- A sponsor of a Class I or Class II device, who is otherwise exempt from submitting a 510(k), would need to submit a 510(k) because the presence or amount of nanoscale material would result in the device being outside the scope of the limitations of exemption described in the general provisions of the applicable regulations (*see* 21 C.F.R. §§ 862.9-892.9).
- A sponsor should submit a new 510(k) for a modification to a previously cleared device that incorporates the use or increased use of nanoscale materials.
- Institutional Review Boards, investigators, and industry should seek input from the FDA on significant risk/nonsignificant risk decisions regarding investigational devices containing nanoscale materials.

For products not subject to pre-market authorization:

- Issue a notice in the *Federal Register* requesting submission of data and other information addressing the effects on product safety of nanoscale materials in products not subject to pre-market authorization.
- Issue guidance or amend existing guidance to describe what additional or distinct information should be submitted to the FDA or generated with regard to:
  - The use of nanoscale materials in food ingredients for which a GRAS notification is submitted or the reduction of particle size into the nanoscale range for food ingredients for which an earlier notification had been submitted and not objected to by the FDA
  - The use of nanoscale materials in new dietary ingredients
- Issue guidance recommending manufacturers consider whether and how the presence of nanoscale materials affects the manufacturing process.
- Issue guidance describing safety issues that manufacturers should consider to ensure that cosmetics made with nanoscale materials are not adulterated.
- Issue guidance on whether a dietary ingredient modified to include nanoscale materials or include a greater proportion of nanoscale materials would still qualify as a dietary ingredient under 21 U.S.C. Section 321(ff)(1), and when the reduction in size into the nanoscale range of an "old" dietary ingredient might trigger the notification process required for a new dietary ingredient on the basis of the presence or amount of nanoscale materials.

4.7.2.3 Issue: Permissible and Mandatory Labeling

According to the Task Force, because current science does not support a finding that classes of products with nanoscale materials necessarily present greater safety concerns than classes of products without nanoscale materials, the Task Force does not believe there is a basis for saying that, as a general matter, a product containing nanoscale materials must be labeled as such. Therefore the Task Force does not recommend that the FDA require such labeling at this time. Instead, the Task Force recommends that the FDA address on a case-by-case basis whether labeling must or may contain information on the use of nanoscale materials.

4.7.2.4   Issue: The National Environmental Policy Act (NEPA)

The Task Force recommends that the FDA take the following actions:

- Take into account, on a case-by-case basis, whether an FDA-regulated product containing nanoscale materials qualifies for an existing categorical exclusion and whether extraordinary circumstances exist.
- Designate a lead within the FDA to coordinate the FDA's approach to its obligations under NEPA regarding nanotechnology.

4.8   THE NATIONAL INSTITUTE FOR OCCUPATIONAL SAFETY AND HEALTH (NIOSH)

The National Institute for Occupational Safety and Health (NIOSH) is deeply engaged in scientific research and development activities pertinent to nanotechnologies, and has been particularly proactive in identifying research needs and helping to fill them. NIOSH maintains an exceptionally well-designed website devoted exclusively to nanotechnology.*

One of the most useful NIOSH initiatives is the issuance of an updated version of its October 2005 document entitled *Approaches to Safe Nanotechnology: An Information Exchange with NIOSH.* NIOSH intends the document to review what is currently known about nanoparticle toxicity and control but notes that it "is only a starting point." According to NIOSH, the document serves as a request from NIOSH to occupational safety and health practitioners, researchers, product innovators and manufacturers, employers, workers, interest group members, and the general public "to exchange information that will ensure that no worker suffers material impairment of safety or health as nanotechnology develops. Opportunities to provide feedback and information are available throughout this document." The document is available on the Internet at http://www.cdc.gov/niosh/topics/nanotech/safenano/pdfs/approaches_to_safe_nanotechnology.pdf.

A summary of findings and key recommendations includes the following:

- Nanomaterials have the greatest potential to enter the body if they are in the form of nanoparticles, agglomerates of nanoparticles, and particles from nanostructured materials that become airborne or come into contact with the skin.
- Based on results from human and animal studies, airborne nanomaterials can be inhaled and deposited in the respiratory tract; and based on animal studies, nanoparticles can enter the bloodstream, and translocate to other organs.
- Experimental studies in rats have shown that equivalent mass doses of insoluble ultrafine particles (smaller than 100 nm) are more potent than large particles of similar composition in causing pulmonary inflammation and lung tumors in those laboratory animals. Toxicity may be mitigated

---

* See http://www.cdc.niosh.gov/niosh/topics/nanotech.

by surface characteristics and other factors, however. Results from *in vitro* cell culture studies with similar materials generally are supportive of the biological responses observed in animals.

- Cytotoxicity and experimental animal studies have shown that changes in the chemical composition, structure of the molecules, or surface properties of certain nanomaterials can influence their potential toxicity.
- Studies in workers exposed to aerosols of manufactured microscopic (fine) and nanoscale (ultrafine) particles have reported lung function decrements and adverse respiratory symptoms; however, uncertainty exists about the role of ultrafine particles relative to other airborne contaminants (e.g., chemicals, fine particles) in these work environments in causing adverse health effects.
- Engineered nanoparticles whose physical and chemical characteristics are like those of ultrafine particles should be studied to determine if they pose health risks similar to those that have been associated with the ultrafine particles.
- Although insufficient information exists to predict the fire and explosion risk associated with nanoscale powders, nanoscale combustible material could present a higher risk than coarser material with a similar mass concentration given its increased particle surface area and potentially unique properties due to the nanoscale.
- Some nanomaterials may initiate catalytic reactions, depending on their composition and structure, that would not otherwise be anticipated from their chemical composition alone.

Nanomaterial-enabled products such as nanocomposites and surface coatings, and materials comprised of nanostructures such as integrated circuits, are, according to NIOSH, unlikely to pose a risk of exposure during their handling and use. Some of the processes (formulating and applying nanoscale coatings) used in their production may lead to exposure to nanoparticles, however. Processes generating nanomaterials in the gas phase, or using or producing nanomaterials as powders or slurries/suspensions/solutions pose the greatest risk for releasing nanoparticles. Maintenance on production systems (including cleaning and disposal of materials from dust collection systems) is likely to result in exposure to nanoparticles if it involves disturbing deposited nanomaterial.

The following workplace tasks, according to NIOSH, may increase the risk of exposure to nanoparticles:

- Working with nanomaterials in liquid media without adequate protection (e.g., gloves) will increase the risk of skin exposure.
- Working with nanomaterials in liquid during pouring or mixing operations, or where a high degree of agitation is involved, will lead to an increased likelihood of the formation of inhalable and respirable droplets.
- Generating nanoparticles in the gas phase in non-enclosed systems will increase the chances of aerosol release into the workplace.

- Handling nanostructured powders will lead to the possibility of aerosolization.
- Maintenance on equipment and processes used to produce or fabricate nanomaterials or the clean-up of spills or waste material will pose a potential for exposure to workers performing these tasks.
- Cleaning of dust collection systems used to capture nanoparticles can pose a potential for both skin and inhalation exposure.
- Machining, sanding, drilling, or other mechanical disruptions of materials containing nanoparticles can potentially lead to aerosolization of nanomaterials.

4.8.1 EXPOSURE ASSESSMENT AND CHARACTERIZATION

Until more information becomes available on the mechanisms underlying nanoparticle toxicity, NIOSH believes that it is uncertain as to what measurement technique should be used to monitor exposures in the workplace. Current research indicates that mass and bulk chemistry may be less important than particle size and shape, surface area, and surface chemistry (or activity) for nanostructured materials. Many of the sampling techniques available for measuring airborne nanoaerosols vary in complexity but can provide useful information for evaluating occupational exposures with respect to particle size, mass, surface area, number concentration, composition, and surface. Unfortunately, presently relatively few of these techniques are readily applicable to routine exposure monitoring.

Regardless of the metric or measurement method used for evaluating nanoaerosol exposures, NIOSH believes that it is critical that background nanoaerosol measurements be conducted before the production, processing, or handling of the nanomaterial/nanoparticle. When feasible, personal sampling is preferred to ensure an accurate representation of the worker's exposure, whereas area sampling (e.g., size-fractionated aerosol samples) and real-time (direct reading) exposure measurements may be more useful for evaluating the need for improvement of engineering controls and work practices.

4.8.2 PRECAUTIONARY MEASURES

Given the limited amount of information about the health risks, NIOSH urges caution to minimize worker exposures. For most processes and job tasks, the control of airborne exposure to nanoaerosols can be accomplished using a wide variety of engineering control techniques similar to those used in reducing exposure to general aerosols. The implementation of a risk management program in workplaces where exposure to nanomaterials exists can help minimize the potential for exposure to nanoaerosols. Elements of such a program should include, according to NIOSH:

- Evaluating the hazard posed by the nanomaterial based on available physical and chemical property data and toxicology or health effects data.
- Assessing potential worker exposure to determine the degree of risk.
- The education and training of workers in the proper handling of nanomaterials (e.g., good work practices).

- The establishment of criteria and procedures for installing and evaluating engineering controls (e.g., exhaust, ventilation) at locations where exposure to nanoparticles might occur.
- The development of procedures for determining the need and selection of personal protective equipment (e.g., clothing, gloves, respirators).
- The systematic evaluation of exposures to ensure that control measures are working properly and that workers are being provided the appropriate personal protective equipment.
- With respect to control measures, engineering control techniques such as source enclosure (i.e., isolating the generation source from the worker) and local exhaust ventilation systems should be effective for capturing airborne nanoparticles. Current knowledge indicates that a well-designed exhaust ventilation system with a high-efficiency particulate air (HEPA) filter should effectively remove nanoparticles.
- The use of good work practices can help minimize worker exposures to nanomaterials. Examples of good practices include cleaning of work areas using HEPA vacuum pickup and wet wiping methods, preventing the consumption of food or beverages in workplaces where nanomaterials are handled, and providing hand-washing facilities and facilities for showering and changing clothes.
- No guidelines are currently available on the selection of clothing or other apparel (e.g., gloves) for the prevention of dermal exposure to nanoaerosols. Some clothing standards incorporate testing with nanoscale particles and therefore provide some indication of the effectiveness of protective clothing with regard to nanoparticles, however.
- Respirators may be necessary when engineering and administrative controls do not adequately prevent exposures. Currently, there are no specific exposure limits for airborne exposures to engineered nanoparticles, although occupational exposure limits exist for larger particles of similar chemical composition. The decision to use respiratory protection should be based on professional judgment that takes into account toxicity information, exposure measurement data, and the frequency and likelihood of the worker's exposure. Preliminary evidence shows that for respirator filtration media, there is no deviation from the classical single-fiber theory for particulates as small as 2.5 nm in diameter. While this evidence needs confirmation, it is likely that NIOSH-certified respirators will be useful for protecting workers from nanoparticle inhalation when properly selected and fit tested as part of a complete respiratory protection program.

4.8.3 OCCUPATIONAL HEALTH SURVEILLANCE

The unique physical and chemical properties of nanomaterials, the increasing growth of nanotechnology in the workplace, available information about biological and health effects in animals associated with exposures to some types of engineered nanoparticles in laboratory studies, and available information about the occupational health effects of incidental ultrafine particles all underscore the need for medical and hazard surveillance for nanotechnology. NIOSH urges every workplace dealing with nanoparticles, engineered nanomaterials, or other aspects of nanotechnology to consider the need for an occupational health surveillance program. NIOSH is in the process of formulating guidance relevant to occupational health surveillance for nanotechnology.

4.9   THE CONSUMER PRODUCT SAFETY COMMISSION (CPSC)

Despite the reportedly growing number of consumer products enabled by nanotechnology, the Consumer Product Safety Commission (CPSC) has been relatively inactive, at least publicly, on nanotechnology matters. The CPSC is a member of the NNI but has yet to develop any public guidance or other documents with the exception of the CPSC's 2005 issuance of its "Nanomaterial Statement"[52].

According to the CPSC, the introduction of consumer products containing nanomaterials into the marketplace may require unique exposure and risk assessment strategies, and one of the primary data needs will be the identification of the specific nanomaterial in the consumer product. The CPSC has jurisdiction over consumer products used in or around the home, except certain items excluded by statute. Examples of products that are regulated by the CPSC include clothing, hazardous household cleaners and substances, electronic devices, appliances, furnishings, building materials, toys, and other juvenile products. Because the federal statutes and regulations do not require pre-market registration or approval of products, the CPSC typically evaluates a product's potential risk to the public only after a product has been distributed in commerce.

The CPSC Nanomaterial Statement provides that the CPSC can assess the potential safety and health risks of nanomaterials, as with other compounds incorporated into consumer products, under existing statutes, regulations, and guidelines, including the Consumer Product Safety Act (CPSA) and the Federal Hazardous Substances Act (FHSA). Under the CPSA, the CPSC evaluates a consumer product to determine whether it contains a defect that creates a "substantial product hazard" or warrants setting a consumer product safety standard by regulation to prevent or reduce an unreasonable risk. According to the Statement,

> "In the absence of an express regulation, as it does with other consumer products, the staff will look to see whether a defective product composed of or containing nanomaterials creates a substantial risk of injury to the public because of, among other factors, the pattern of the defect, the number of defective products distributed in commerce, and the severity of the risk."

The Statement also notes that manufacturers, retailers, and distributors of nanomaterial products "have the same reporting obligation as those of other products, namely to report to the Commission immediately if they obtain information that reasonably supports the conclusion that such product fails to comply with an applicable consumer product safety rule; contains a defect which could create a substantial product hazard; or creates an unreasonable risk of serious injury or death."

The CPSC assesses a product's potential chronic health effects to consumers under the FHSA, which is risk based and addresses both acute and chronic hazards.

To be considered a "hazardous substance" under the FHSA, a consumer product must satisfy a two-part definition: (1) it must be toxic under the FHSA, and (2) it must have the potential to cause "substantial personal injury during or substantial illness during or as a proximate result of any customary or reasonably foreseeable handling or use." Therefore, according to the Statement, exposure and the subsequent risk must be considered in addition to toxicity when assessing potential hazards under the FHSA. The CPSC assesses chronic toxicity data using the federal regulations summarizing the guidelines for determining chronic toxicity at 16 C.F.R. Section 1500.135. The CPSC Nanomaterial Statement states that the CPSC "is currently reviewing and updating the chronic hazard guidelines to address, among other things, nanomaterial use in consumer products."

The Statement also says:

> "Because of the wide variation in potential health effects and the dearth of data on exposure and toxicity data of specific nanomaterials, CPSC staff is unable to make any general statements about the potential consumer exposures to, or the health effects that may result from exposure to nanomaterials during consumer use and disposal."

According to the CPSC, identifying any potential health hazards from a specific product "will require characterization of the materials to which a consumer is exposed during product use, including assessment of the size distribution of the materials released." The CPSC states that once the exposure has been characterized, "toxicological data that [are] appropriate for the particle sizes represented in the exposure assessment will be used in any assessment of health risks."

The CPSC is involved in federal and private initiatives addressing the production and use of nanomaterials, including the National Science and Technology Council's Subcommittee on Nanoscale Science, Engineering, and Technology and Interagency Working Group on Nanotechnology Environmental and Health Implications, the American National Standards Institute, ASTM International, and the International Life Sciences Institute. The activities of these groups, according to the CPSC Nanomaterial Statement, include promoting responsible research and development of nanomaterials that can be used in consumer products and providing information on new products that are being introduced into the market. The Statement is available on the Internet at http://www.cpsc.gov/library/cpscnanostatement.pdf.

4.10   EMERGING STATE AND LOCAL REGULATION OF NANOMATERIALS

4.10.1   CITY OF BERKELEY ORDINANCE

With federal regulations specific to engineered nanoscale materials neither in place nor expected anytime soon, states and municipalities are starting to assert jurisdiction over the nanotechnology industry. On December 12, 2006, the Berkeley, California, City Council, acting on the recommendation of the city's Community Environmental Advisory Commission (CEAC), unanimously adopted an ordinance that requires businesses to report nanoparticles being used, provide available toxicological information, and outline measures for safe handling of the materials. Under the municipal ordinance, believed to be the first of its kind in the United States, all facilities that manufacture or use manufactured nanoparticles must submit a written report of the current toxicology of the nanoscale materials reported and how the facility will safely handle, monitor, contain, dispose, track inventory, prevent releases, and mitigate such materials. The ordinance is set forth in Title 15 of the Berkeley Municipal Code, which requires the filing of disclosure information for hazardous materials when certain quantities are exceeded [53].

According to the formal CEAC Recommendation, questions about the need for the city to implement a nanoparticle reporting requirement arose during the design phase of the molecular foundries at the University of California and Lawrence Berkeley Lab; both institutions had indicated that they lacked special knowledge or tools to manage nanoparticles [54]. After consideration and input from others, including the EPA, "the recommended self-reporting was considered to be a minimum regulation for nanotechnology facilities" [54].

The CEAC Recommendation notes that, in many cases, businesses "will not find sufficient information to determine the health impacts of a material. In such cases, it is hoped that a precautionary approach [will] be used when handling the materials" [54]. It further states:

"Nanoparticles behave differently [than] macro-particle compounds and should be handled and mitigated differently. Handlers may not know much about the materials they are handling, as new information is published, the handlers should keep updating their knowledge, since government is not doing a good job regulating these materials."

Finally, although the no action alternative was considered by the CEAC, "clearly no action has potentially unacceptable consequences for nanoparticle workers and the community" [54].

4.10.2   Cambridge, Massachusetts, Ordinance

According to recent press reports, the Cambridge, Massachusetts, City Council is contemplating becoming the second municipality in the country to regulate nanotechnology [55]. In a recent vote, the council requested that the city's public health department undertake a study of the Berkeley ordinance and assess whether a similar enactment made sense for Cambridge, which is home to quite a few nanotechnology entities.

4.11   PRIVATE NANOTECHNOLOGY STEWARDSHIP INITIATIVES

Environmental Defense (ED) and DuPont formally announced in June 2007 the release of their *Nano Risk Framework*, which defines "a systematic and disciplined process for identifying, managing, and reducing potential environmental, health, and safety risks of engineered nanomaterials across all stages of a product's 'lifecycle'

— its full life from initial sourcing through manufacture, use, disposal or recycling, and ultimate fate."*

ED and DuPont began their collaborative effort to develop the *Framework* in September 2005. They released a draft version to the public on February 26, 2007, and received comments from a diverse array of stakeholders — government, academia, public interest groups, and both large and small companies. In addition to considering the various comments, ED and DuPont conducted pilot-testing on surface-treated high-rutile phase titanium dioxide (TiO_2), single- and multi-walled carbon nanotubes (CNTs), and nano-sized zero-valent iron (nano-Fe^0) "to ensure that [the *Framework*] is flexible, practical, affordable, and effective." The final document issued "offers guidance on the key questions an organization should consider in developing applications of nanomaterials, and on the information needed to make sound risk evaluations and risk-management decisions." The *Framework* is intended to support ongoing regulatory initiatives — not replace them.

ED and DuPont believe that the *Framework*, which is aimed primarily at organizations, both private and public, that are actively working with nanomaterials and developing associated products and applications, will help users organize and evaluate currently available information; assess, prioritize, and address data needs; and communicate clearly how risks are being mitigated. Ultimately, ED and DuPont "believe that the adoption of the *Framework* can promote responsible development of nanotechnology products, facilitate public acceptance, and support the formulation of a practical model for reasonable government policy on nanotechnology safety." Further information on the Framework is available in Chapter 11, *infra*.

4.12   INTERNATIONAL DEVELOPMENTS

Globally, there is much activity regarding the regulation of nanotechnology. It is not the intent of this section to address these initiatives in detail. Rather, the intent is to identify a few of the more prominent developments of which readers should be aware.

4.12.1   REGISTRATION, EVALUATION, AUTHORIZATION, AND RESTRICTION OF CHEMICALS (REACH)

In the European Union (EU), the enactment in June 2007 of the Registration, Evaluation, Authorization, and Restriction of Chemicals (REACH) began what will eventually result in the most sweeping change in chemical management policy ever. REACH is the EU's complex new chemical management regulation.** While no REACH provision explicitly cites nanoscale material, it is widely believed that REACH includes nanoscale materials. REACH is complicated. In total, it encompasses more than 140 different articles, 17 distinct annexes, almost 300 pages of

---

* A complete copy of the *Framework* and other related information are available at http://nanorisk-framework.com/page.cfm?tagID=1095.

** REACH is available at http://eur-lex.europa.eu/LexUriServ/site/en/oj/2007/l_136/l_ 13620070529en 00030280.pdf. The regulation entered into force on June 1, 2007, although most of its key provisions will not apply until June 1, 2008. See REACH Art. 141.

(reformatted) text, and hundreds of pages of guidance, with the latter figure expected to grow considerably as more guidance is issued.

The core of REACH is its registration requirement, which mandates that all chemicals manufactured or imported into the EU in quantities of 1 metric ton or more per year be registered with the newly created European Chemicals Agency (ECHA).* The registration obligation applies to legal entities — all manufacturers and importers of "substances" on their own or in "preparations" (i.e., mixtures), and all producers and importers of "articles" meeting certain criteria — that are established within the EU and that meet the 1-metric-ton-per-year threshold.**

Registration will entail the generation of substance-specific health and safety data; preparation of a technical dossier; and for those substances manufactured or imported in quantities of 10 metric tons or more per year, an assessment of the risks posed by the substance, including relevant exposure scenarios, and the development and communication of appropriate risk management measures [56].

For "phase-in substances," which include existing chemicals,*** the registration process will proceed in phases. To benefit from the extended registration deadlines — 3½, 6, and 11 years from June 1, 2007, depending on the annual volume and hazard of the substance — manufacturers and importers (as well as producers and importers of certain articles) must pre-register their substances between June 1, 2008, and December 1, 2008.**** Pre-registration will enable a company to continue manufacturing or importing the substance until the extended registration deadline is reached.*****

For most entities, the initial step under REACH will be the pre-registration of phase-in substances. Pre-registration will entail the electronic submission to the ECHA of certain basic information on the chemical and the pre-registrant [57, 58]. By January 1, 2009, the ECHA intends to publish on its website a list of the pre-registered substances. The list will not identify the pre-registrants, but this information will be available to all companies that have pre-registered the same substance [58]. The companies will then be required to participate in a Substance Information Exchange Forum (SIEF) for the substance, with the aim of the SIEF being to facilitate the sharing of existing data on the chemical, the collective identification of data gaps, and cost-sharing with respect to the generation of any new data [59, 60].

---

* See REACH Art. 5-7. Under Article 5, the non-registration of a substance that is required to be registered means that the substance cannot be manufactured, imported, or otherwise placed on the EU market. This REACH principle is often referred to as the "no data, no market" principle.

** See REACH Art. 3, 6-7; see generally ECHA, Guidance on Registration (June 2007) at 19–21, available at http://reach.jrc.it/03_rdds_web_content/ registration_en/registration_en.pdf. Definitions of the key REACH terms appear in Article 3 of the regulation.

*** The term "phase-in substance" is defined in REACH Article 3(20).

**** See REACH Art. 23, 28; see generally Guidance on Registration at 41–44, 52. Final guidance on the pre-registration process is expected soon.

***** See REACH Art. 21(1), 23(1)–(3); see generally Guidance on Registration at 52; ECHA, Guidance on Data Sharing (Sept. 2007) at 20, available at http://reach.jrc.it/docs/guidance_document/data_sharing_en.pdf. Pre-registration is not required; but for a phase-in substance that is not pre-registered, the company cannot legally manufacture and/or import the substance subsequent to June 1, 2008, until 3 weeks after it has submitted a complete registration. See Guidance on Registration at 52; Guidance on Data Sharing at 23.

Unless they import the chemicals they utilize, downstream users (DU) of chemicals (e.g., formulators of mixtures, users of chemicals in industrial activities) do not have registration obligations under REACH. Each DU, however, must identify, apply, and, where suitable, recommend appropriate risk management measures, and may have certain risk assessment, communication, and notification obligations under the regulation.*

Under the REACH authorization provisions, "very high concern" chemicals will be included on a list of candidate chemicals that the ECHA is expected to publish in late 2008 [61]. Eventually, approximately 1500 "substances of very high concern" (SVHCs) are expected to be taken from the candidate list and included in REACH Annex XIV, the list of substances that will be subject to REACH authorization. Once included in Annex XIV, authorization from the European Commission (EC) will be needed before the substance can be marketed or used.** An application for authorization must include an analysis of alternatives and a substitution plan where a suitable alternative exists [62]. Thus, the REACH authorization system is designed to "assur[e] that ... [SVHCs] are progressively replaced by suitable alternative substances or technologies where these are economically and technically viable" [63]. It is widely anticipated that some, perhaps many, of the manufacturers of SVHCs will cease manufacturing them, forcing DUs either to reformulate their products or cease producing those products in the absence of viable substitutes. Some uses of chemicals, moreover, may be limited under the REACH restriction provisions, and authorization for those uses would not be granted.

As the EU authorities have explained, "REACH is very wide in its scope" [64] and "applies to all substances with a few exemptions" [65]. Exemptions from all aspects of REACH exist for radioactive substances, substances under customs supervision, non-isolated intermediates, the transport of dangerous substances, and waste.*** There is no explicit exemption for nanoscale materials. Partial exemptions exist for, *inter alia*, substances listed in Annex IV (minimum risk substances) or covered by Annex V (e.g., incidental reaction products, byproducts, natural substances) and re-imported substances [66, 67], and "[a] number of other substances are exempted from parts of the provisions of REACH, where other equivalent legislation applies" [65, 68].

Although nano-specific provisions in REACH were explicitly considered when developing REACH, they were ultimately rejected. As noted, however, there is no question that REACH applies to nanoscale substances. What is less clear is whether the ECHA will consider each nanoscale substance as equivalent to the macroscale version. If so, registrants would be required to develop the necessary data set required under Article 10 for the dossier and provide guidance on the safe use of these materials as a condition of pre-market approval [69].

* See REACH Art. 31-39; see generally ECHA, "Downstream Users," available at http://reach.jrc.it/downstream_users_en.htm. Note that only EU entities are DUs under REACH. See REACH Art. 3(13)

** See REACH Art. 56. Annex XIV will specify for each SVHC a date after which the placement on the market and use of the substance will be prohibited unless an authorization is granted. See REACH Art. 58(1)(c)(i).

*** See REACH Art. 2(1)-2(2); see generally ECHA, Guidance for the Navigator (June 2007) at 7–9, available at http://reach.jrc.it/03_rdds_web_content/navigator_en/navigator_en.pdf. Given that all aspects of the regulation are inapplicable, it is probably more accurate to state that these substances are excluded from REACH.

4.12.2   THE ORGANIZATION FOR ECONOMIC COOPERATION AND DEVELOPMENT (OECD) INITIATIVES

Another very important international initiative includes those activities sponsored by the Organization for Economic Cooperation and Development (OECD). The OECD is an intergovernmental organization in which representatives from 30 industrialized countries in North America, Europe, Asia, and Pacific regions, and the European Commission coordinate and harmonize polices, discuss issues of shared concern, and work together to address international problems. Two OECD groups are especially pertinent to nanotechnology. In 2006, the OECD Council established the Working Party on Manufactured Nanomaterials (WPMN) as a subsidiary of the Chemicals Committee. The WPMN is working on a series of six specific projects involving nanotechnology:

1. Database on human health and environmental safety research
2. EHS research strategies on manufactured nanomaterials
3. Safety testing of a representative set of manufactured nanomaterials
4. Manufactured nanomaterials and test guidelines
5. Cooperation on voluntary schemes and regulatory programs
6. Cooperation on risk assessments and exposure measures

Additionally, in 2006, the OECD Committee for Scientific and Technological Policy (CSTP) created a Working Party on Nanotechnology (WPN). The WPN is intended to provide advice and comment on emerging policy-relevant issues of science, technology, and innovation related to the responsible development of nanotechnology.

4.12.3   CANADIAN INITIATIVES

Environment Canada and Health Canada have developed a proposal for a regulatory framework for nanomaterials under the Canadian Environmental Protection Act, 1999 (CEPA). Under the CEPA, the Ministers of the Environment and Health must conduct environmental and human health risk assessments and manage appropriately any risks arising from industrial chemical substances entering the Canadian market. Stakeholders from industry, non-governmental organizations, and other interested parties provided feedback on the proposed approach to developing a regulatory framework, as well the options for gathering information on industrial nanomaterials as part of the first phase of the program.

According to Environment Canada and Health Canada, a regulatory framework for nanomaterials needs to be developed in a way that is scientifically robust and harmonizes the outcomes of international efforts. Environment Canada and Health Canada have proposed the development of a regulatory framework for nanomaterials consisting of two phases of implementation based on shorter and longer-term objectives.

Phase 1 (started Fall 2006):

a. Continue work with international partners (OECD, International Organization for Standardization [ISO]) to develop scientific and research capacities.
b. Inform potential notifiers of their regulatory responsibilities under the current framework.
c. Develop initiatives to gather information from industry on the uses, properties, and effects of nanomaterials.
d. Consider whether amendments to CEPA or the New Substances Notification Regulations (NSNR) would be needed to facilitate the risk assessment and management of nanomaterials.

Phase 2 (starting 2008):

a. Resolution of terminology and nomenclature by ISO TC229.
b. Consider establishing data requirements under the NSNR specific to nanomaterials.
c. Consider the use of the Significant New Activity (SNAc) provision of CEPA to require notification of nanoscale forms of substances already on the Domestic Substances List (DSL).

For more information on the proposal, see *Proposed Regulatory Framework for Nanomaterials under the Canadian Environmental Protection Act, 1999.*

REFERENCES

1. Toxic Substances Control Act. 2006. 15 U.S.C., §§ 2601-2692.
2. Toxic Substances Control Act. 2006. § 2(b), 15 U.S.C., § 2601(b)(2).
3. Toxic Substances Control Act. 2006. § 5(a), 15 U.S.C., § 2604(a).
4. Toxic Substances Control Act. 2006. § 8(b)(1), 15 U.S.C., § 2607(B)(b)(1).
5. Toxic Substances Control Act. 1979. Availability of TSCA Initial Inventory; Beginning of 210-Day Reporting Period for Revised Inventory, 44 *Fed. Reg.,* 28,558 (15 May).
6. Toxic Substances Control Act. 2007. § 3(9), 15 U.S.C., § 2602(9).
7. U.S. Environmental Protection Agency. 2007. 40 C.F.R., §§ 710.3, 720.3(v), 720.25(a).
8. U.S. Environmental Protection Agency. TSCA Chemical Substance Inventory. http://www.epa.gov/opptintr/newchems/pubs/invntory.htm.
9. U.S. Environmental Protection Agency. 2007. 40 C.F.R., § 720.25(a)(5).
10. U.S. Environmental Protection Agency. 2007. EPA Form 7710-25. 40 C.F.R., § 720.25(a)(5) (2007). Pt. 720, subpt. C. http://www.epa.gov/opptintr/newchems/pubs/pmnforms.htm.
11. Toxic Substances Control Act. 2006. § 5(h)(3), 15 U.S.C., § 2604(h)(3).
12. U.S. Environmental Protection Agency. 2007. 40 C.F.R., § 720.36.
13. U.S. Environmental Protection Agency. 2007. 40 C.F.R., § 720.3(cc).
14. U.S. Environmental Protection Agency. 1986. New Chemical Information Bulletin: Exemptions for Research and Development and Test Marketing (November): 5. http://www.epa.gov/opptintr/newchems/pubs/tmeranddbulletin.pdf.
15. U.S. Environmental Protection Agency. 2007. 40 C.F.R., § 723.50(c)(2).
16. 70 *Fed. Reg.,* 24574 (May 10, 2005).

* See http://www.ec.gc.ca/substances/nsb/eng/nano_e.shtml.

17. U.S. Environmental Protection Agency. 2005. Overview Document on Nanoscale Materials. National Pollution Prevention and Toxics Advisory Committee (22 November): 6. http://www.epa.gov/opptintr/npptac/pubs/nanowgoverviewdocument20051125. pdf.

18. Bergeson, L. 2005. Nanotechnology and TSCA, 6(3) *ABA Pesticide, Chemical Regulation and Right-to-Know Committee Newsletter* 11 (April). http://www.abanet.org/environ/committees/pesticides/newsletter/apr05/pesticides0405.pdf.

19. Gulliford, J.B., Assistant Administrator for Prevention, Pesticides & Toxic Substances. 2006. Letter to Stakeholders (18 October). http://www.epa.gov/oppt/nano/nano-letter.pdf.

20. 72 *Fed. Reg.,* 38081.

21. U.S. Environmental Protection Agency. 2007. Concept Paper for the Nanoscale Materials Stewardship Program under TSCA. (July): 1–6, 13. http://epa.gov/oppt/nano/nmsp-conceptpaper.pdf. (Accessed September 10, 2007)

22. U.K. Department for Environment, Food and Rural Affairs (Defra). 2006. UK Voluntary Reporting Scheme for engineered nanoscale materials. (September): 3. http://www.defra.gov.uk/environment/nanotech/policy/pdf/vrs-nanoscale.pdf.

23. U.K. Department for Environment, Food and Rural Affairs (Defra). 2006. Consultation on a Proposed Voluntary Reporting Scheme for Engineered Nanoscale Materials. (March): 3. http://www.defra.gov.uk/corporate/consult/nanotech-vrs/consultation.pdf.

24. U.S. Environmental Protection Agency. 2007. Nanotechnology White Paper. EPA/100/B-07/001. Prepared for the U.S. Environmental Protection Agency by members of the Nanotechnology Workgroup, a group of EPA's Science Policy Council (February): 2; 3, 19-20, 25, 66, 67. http://www.epa.gov/ncer/nano/publications/whitepaper12022005.pdf and http://www.epa.gov/osa/nanotech.htm (accessed September 10, 2007).

25. 71 *Fed. Reg.,* 14205 (Mar. 21, 2006).

26. 72 *Fed. Reg.,* 7435.

27. 7 *U.S.C., § 136 et seq.*

28. 72 *Fed. Reg.,* 54039.

29. 42 U.S.C., § 7408(a)(2).

30. 42 U.S.C., § 7409(b)(1).

31. *Whitman v. American Trucking Associations,* 531 U.S. 457 (2001).

32. *Lead Industries Association. v. EPA,* 647 F.2d 1130 (D.C. Cir.), *cert denied,* 449 U.S. 1042 (1980).

33. *American Petroleum Institute v. Costle,* 665 F.2d 1176 (D.C. Cir. 1981).

34. U.S. Environmental Protection Agency. 2005. White Paper: Preliminary Analysis of Proposed $PM_{2.5}$ NAAQS Alternatives (20 December). http://www.epa.gov/air/particles/pdfs/whitepaper20051220.pdf.

35. 42 U.S.C., § 7412.

36. CAA, § 112(b)(2), 42 U.S.C. § 7412(b)(2).

37. CAA, § 112(b)(3)(B), 42 U.S.C. § 7412(b)(3)(B).

38. CAA, § 112(d)(2), 42 U.S.C. § 7412(d)(2).

39. 42 U.S.C. § 7412(d)(5).

40. CAA § 112(f)(2), 42 U.S.C. § 7412(f)(2).

41. 42 U.S.C. § 7412(r)(1).

42. CAA § 112(r)(2)(A), 42 U.S.C. § 7412(r)(2(A).

43. CAA § 112(r)(3)-(5), 42 U.S.C. § 7412(r)(3)-(5).

44. CWA § 101(a), 33 U.S.C. § 1251(a).

45. CWA § 301(a), 33 U.S.C. § 1311(a).

46. U.S. Environmental Protection Agency. 40 C.F.R. §§ 122.2, 122.3(c).

47. U.S. Environmental Protection Agency. 40 C.F.R. Part 403.

48. U.S. Environmental Protection Agency. 40 C.F.R. § 261.3(c)(2)(i).

49. Woodrow Wilson International Center for Scholars. Nanotechnology Consumer Products Inventory: Project on Emerging Nanotechnology. http://www.nanotechproject. org/inventories.

50. U.S. Environmental Protection Agency. 40 C.F.R. § 260.10.

51. U.S. Environmental Protection Agency. 40 C.F.R. § 263.10(a).

52. Consumer Product Safety Commission. 2005. Nanomaterial Statement. http://www. epa.gov/opptintr/nano/scope.htm.

53. Berkeley Municipal Code. 2006. Title 15 Hazardous Materials. http://www.ci.berkeley. ca.us/citycouncil/2006citycouncil/packet/121206/2006-12-12Item03-Ord-Nanoparticles.pdf.

54. Community Environmental Advisory Commission (CEAC). (undated) Memorandum to Honorable Mayor and Members of the City Council (CEAC Recommendation), p. 2-3.   http://www.ci.berkeley.ca.us/citycouncil/2006citycouncil/packet/120506/2006-12-05%20Item%2013%20Manufactured%20Nanoparticle%20Health%20and%20Safety%20Disclosure.pdf.

55. Cambridge Considers Nanotech Curbs, Boston Globe (Jan. 26, 2007), *available at* http://www.smalltimes.com/articles/article_display.cfm?Section=ONART&C=Mater &ARTICLE_ID=283218&p=109.

56. ECHA. 2007. Guidance on Registration (June): 12–13. http://reach.jrc.it/03_rdds_web_ content/ registration_en/registration_en.pdf.

57. Registration, Evaluation, Authorization, and Restriction of Chemicals (REACH). Art. 28(1). http://eur-lex.europa.eu/LexUriServ/site/en/oj/2007/l_136/l_13620070529en000 30280.pdf.

58. ECHA. 2007. Guidance on Data Sharing (September): 20, 31–32. http://reach.jrc. it/docs/guidance_document/data_sharing_en.pdf.

59. REACH. Art. 29. http://eur-lex.europa.eu/LexUriServ/site/en/oj/2007/l_136/l_136200 70529en00030280.pdf.

60. ECHA. Data-Sharing. http://reach.jrc.it/data_sharing_en.htm.

61. ECHA. Authorizations. http://reach.jrc.it/authorisation_en.htm.

62. REACH. Art. 62(4)(e)-(f). http://eur-lex.europa.eu/LexUriServ/site/en/oj/2007/l_136/ l_ 13620070529en00030280.pdf.

63. REACH. Art. 55. http://eur-lex.europa.eu/LexUriServ/site/en/oj/2007/l_136/l_136200 70529en00030280.pdf.

64. EC. 2007. REACH in Brief (October): 6. http://ec.europa.eu/environment/chemicals/ reach/pdf/2007_02_reach_in_brief.pdf.

65. ECHA. Chemicals Covered. http://reach.jrc.it/chemicals_covered_en.htm.

66. REACH. Art. 2(7). http://eur-lex.europa.eu/LexUriServ/site/en/oj/2007/l_136/l_136200 70529en00030280.pdf.

67. ECHA. 2007. Guidance for the Navigator (June): 7-9. http://reach.jrc.it/03_rdds_ web_content/navigator_en/navigator_en.pdf.

68. REACH. Art. 2(5)-2(6). http://eur-lex.europa.eu/LexUriServ/site/en/oj/2007/l_136/ l_13620070529en00030280.pdf.

69. Bowman, D. and G. van Calster. 2007. Reflecting on REACH: Global Implications of the European Union's Chemicals Regulation. In *Nanotechnology Law & Business*, 4(3): 375–383.

5 Analyses of Nanoparticles in the Environment

Marilyn Hoyt
AMEC Earth & Environmental

CONTENTS

The rapid explosion of production and use of engineered nanoparticles has outpaced the scientific community's ability to monitor their presence in the environment. Without measurement data, it is not possible to fully evaluate whether the promises of nanoparticles are accompanied by significant ecological or human health risks. Numerous national and international agencies and research groups have recognized this gap and put in place research programs to address it. However, the technical requirements for the detection and characterization of nanoparticles in complex environmental systems push the limits of current sampling techniques and instrumentation. In most cases, multiple complementary measurements are likely necessary to detect and understand the importance of nanoparticles in air, water, or soil because physical properties as well as chemical composition determine activity and environmental impact or risk. Environmental analyses of nanoparticles are not common offerings at commercial environmental laboratories at this time, and they are not likely to become so in the near future.

In the manufacturing industry, the development and production of nanoparticle materials for commercial applications are supported by an array of analytical methods. While numerous methods can successfully characterize the chemistry and physical properties of nanoparticles in relatively pure states and under defined conditions, the applicability of these methods to nanoparticles in environmental settings may be more limited. Once nanoparticles enter the environment, they may cluster to form larger particles, interact with particles from natural sources, or change chemically. Conventional environmental analysis methods as developed and standardized by the U.S. Environmental Protection Agency (EPA) are bulk analyses; they can detect the primary chemical constituents of nanoparticle materials but little else of use for characterizing risk from them. In addition, the target nanoparticles may only be a minor component of an environmental sample and fall below the detection limits of standard EPA chemical analysis methods. Collection and separation of nanoparticles from larger environmental particles, when even possible, are difficult, and their analysis is in most cases time-consuming and costly. No standard methods with prescribed quality control requirements for environmental nanoparticle analyses exist, and only limited traceable standards have been developed.

Aside from the technical challenges to nanoparticle measurement in environmental media, the lack of specific regulations limits the incentive for commercial environmental laboratories to put in place the costly instrumentation and the high degree of expertise that will be required to offer nanoparticle analyses to government, private industry, or public groups. While there is some concern for possible environmental risks from nanoparticles, manufacturers, users, and site owners currently are not required to address these concerns with actual environmental measurement data. As a result, most technical advances and data that do exist for environmental analyses have come from academic laboratories and governmental or privately funded research laboratories. The applicability of regulatory statutes as discussed in Chapter 4 of this book continues to be debated. The Toxic Substances Control Act (TSCA), the Clean Water and Clean Air Acts (CWA, CAA), the Resource Conservation and Recovery Act (RCRA), and the Federal Insecticide, Fungicide, and Rodenticide Act (FIFRA) drove method development for numerous industrial chemicals in the environment. Regulatory requirements applicable to nanomaterials likewise would be expected to drive the development and standardization of environmental nanoparticle analytical methods for wider application, as well as to foster competition in an emerging market for laboratory services. Instrumentation and staffing costs will, however, remain a barrier to entry into the field for most commercial laboratories currently offering environmental services.

5.1 ANALYTICAL METHODS

The production of nanoparticle materials typically requires control of the chemical composition, size, shape, and surface characteristics of the material. Many of the analytical techniques applied for the analysis of nanoparticles during development and production also are critical to laboratory studies of fate and transport and exposure effects to ensure that the material being tested is fully understood. These methods also may be components of analyses to detect nanoparticles after their release into the environment, dispersion in air or water, or uptake into organisms [1].

This chapter discusses highlights of the most widely used techniques, providing the basic science of the analyses and describing the type of information that can be expected and reported for possible environmental applications. These techniques, as listed in Table 5.1, represent what must be considered initial approaches of researchers to address environmental issues; it is likely that over time, other current techniques or newly developed instrumentation will also prove useful. Representative citations are provided where methods have proven successful for analyses of nanoparticles present in air, water, or soils. However, it should be noted that most environmental analyses reported to date for nanoparticles have focused on natural species such as colloids in water or on combustion-related emissions. Engineered nanoparticles have been characterized in laboratory studies and in indoor air monitoring programs, but only limited studies designed to detect their releases into or fate in ambient air, surface or ground waters, or soils or waste have been reported [2].

More in-depth discussions of the theoretical basis for each measurement technique, specifics for instrument design, detection options, and data examples can be found in a review article [3] that discusses more than 30 measurement techniques in detail, presenting the theory and advantages and limitations to each. Laboratory analyses, real-time methods, and portable instrumentation for particulate characterization from mobile source emissions are reviewed in a literature survey for the California Air Research Board (ARB) [4]. Many of the methods discussed and equipment illustrated are also potentially applicable to measurement of nanoparticles from other sources in the environment. A recent U.S. EPA symposium on nanoparticles in the environment discussed the challenges involved, and also presented highlights of applicable measurement methods [5].

5.1.1 NANOPARTICLE IMAGING: SIZE, SHAPE, AND CHEMICAL COMPOSITION

5.1.1.1 Electron Microscopy

Electron microscopy is comparable to light microscopy, except that a beam of electrons rather than light is used to form images. Electron beams have a much shorter wavelength than light and, as a result, they can provide the resolution required to

TABLE 5.1
Methods for Environmental Analyses of Nanoparticles

Technique	Parameters Measured	Resolution/Sensitivity	Limitations/Advantages	Environmental Applications
Nanoparticle Imaging				
Electron microscopy (SEM, TEM, ESEM)	Particle size, shape, texture, crystalline vs. amorphous structure, elemental composition, bonding	1 nm SEM, <0.1 nm TEM	Particle-by-particle analysis, time-consuming. Sample preparation. High vacuum for SEM, TEM may alter particles. ESEM allows imaging in water or other liquid media	Ambient air studies [11], nanoparticle characterization for laboratory studies of fate, toxicity [7–10]
Scanning probe microscopy (STM, AFM)	Particle size, morphology	0.5 nm	Particle-by-particle analysis. Analysis at ambient pressure, particles may be in solution	Ambient air studies, natural colloids [15–17, 20, 21]
Compositional Analysis				
Single-particle mass spectrometry	Chemical composition, organic and inorganic species	3 nm particle	Continuous analysis of particles in air stream	Atmospheric studies, vehicular emissions [23, 24]
Particle-induced x-ray (PIXE)	Elemental mapping of nanofilms or collected nanoparticles	1 micron	Requires radioactive source.	Air pollution studies [28]
Surface Area				
BET	Average surface area on a mass basis	2000 m²/g	Laboratory-based instrument; requires relatively pure bulk sample of chemically homogenous material.	Characterization for laboratory studies of fate, toxicity [29]
Epiphaniometer	Active surface area	10–20 nm particles, 0.003 m²/cm³	Requires radioactive lead source	Ambient air studies [30]
Aerosol diffusion charger	Aerosol surface area	10 to 100 nm in diameter	Fast response	Ambient air [31]

Size Distribution

Technique	Description	Size	Comments	Applications/References
Electrostatic classifier (DMA, NDMA, DMPS, SMPS)	Particle distribution based on assumed spherical shape	5 nm	Monitors on real-time basis; size will not necessarily be same as from imaging technique	Releases during nanopowder use [33]
Cascade impactor, MOUDI	Particle distribution based on aerodynamic diameter	<30 nm diameter <10 nm (MOUDI)	Time-integrated average distributions; particles collected may be analyzed subsequently by microscopy	Ambient air studies, vehicle emissions [35]
Electrical impactor (ELPI)	Particle distribution based on aerodynamic diameter	7 nm, >90 nanoparticles/cm^3 air; 5 ng/m3	Real-time particle counts	Indoor air, ambient air studies, vehicular emissions [36, 37]
Light scattering (DLS, PLS, QELS)	Particle size based on hydrodynamic diameter	0.7 nm	In situ measurements possible	Characterization of nanomaterials prior to laboratory studies [38–40]

Particle Concentration/Surface Area in Air

Technique	Description	Size	Comments	Applications/References
Condensation particle counter	Particle concentration in air stream	3 nm	No information on particle size, shape composition. Hand-held units available, real-time data.	Indoor air monitoring, worker exposure studies [43]
Electrical aerosol detector	Aerosol diameter concentration, calculated from a number concentration multiplied by average diameter	10 nm	Real-time data generation, field-portable instrumentation	Ambient air studies [45]

Particles in Aqueous Samples

Technique	Description	Size	Comments	Applications/References
Field-Flow Fractionation	Particle separation by size	1 nm diameter; 1–5000 ng/L for elemental composition	Must be combined with subsequent analysis to assess size, (e.g., DLS). Can combine with ICPMS, ESEM.	Natural colloids, iron oxide/hydroxide colloids [49, 50]

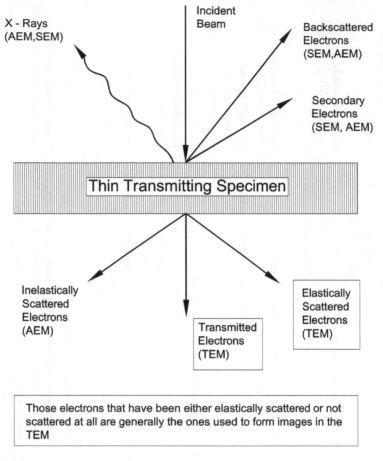

FIGURE 5.1 Electron microscopy. (From J. Mansfield, University of Michigan. With permission.)

form clear images of nanomaterials. There are two major types of electron microscopy: (1) transmission electron microscopy (TEM) and (2) scanning electron microscopy (SEM). As a beam of electrons hits the surface of a particle or film, electrons can be deflected off the surface or, in collisions with atoms of the material, release light, knock off secondary electrons from atoms in the material, or cause the emission of x-rays. Some electrons also pass through the material, either directly or with some scattering due to collisions with the particle atoms (Figure 5.1).

With SEM, emissions from the top of a surface impacted by the electron beam are detected and measured. A variety of instruments can be used to detect the backscattered electrons, secondary electrons, x-rays, or light generated above the surface. Each detector adds its own acronym to the analysis technique (e.g., EDS [energy dispersive x-ray spectroscopy], EDX [energy dispersive x-ray], and XEDS [x-ray energy dispersive spectroscopy] all refer to x-ray detection techniques that provide structural or chemical composition information when paired with SEM). Auger electron microscopy or spectroscopy (AEM or AES), which measures the energy of ejected electrons, also is useful for elemental composition information. Paired with these different detectors, SEM can provide information on the size and shape of a particle, three-dimensional topographic information on surface features and texture, crystalline or amorphous structure, and elemental composition. The technique is most useful for measurements of particles in the range of 50 nanometers (nm) or higher, although stronger electron sources can achieve spatial resolution of 1 nm. More advanced detectors are available now that can charactize the difference in chemistry between the top 2 nm of a particle and its interior.

With TEM, the measurements are taken underneath the material. The portion of the electron beam that passes through the particle can be projected onto a fluorescent screen to form a two-dimensional image of the particle. Resolution of less than 0.1 nm can be achieved, making it a primary tool for characterization of the smallest nanoparticles. As with SEM, a variety of detectors can be used to detect scattered electrons and x-rays released by the interactions of the electron beam with the atoms of the particles. TEM analyses can be designed to determine the elemental composition of the particle and the chemical bonding environment, particle shape and size, and its crystalline or amorphous structure. TEM also can be conducted in a scanning mode (STEM), where the narrowly focused electron beam scans over the particle for maximum sensitivity and resolution. A more detailed introduction to TEM is available on the Internet [6].

Researchers frequently use SEM and TEM to characterize nanoparticles before their use in laboratory experiments and to monitor progress or results. TEM has been used to characterize TiO_2 and fullerene for inhalation and aquatic toxicity studies [7, 8]. Rothen-Rutishauser et al. [9] used TEM techniques to visualize TiO_2 and gold nanoparticles absorbed into red blood cells; and Sipzner et al. [10] monitored the dermal absorption of TiO_2 nanoparticles using TEM.

Reported environmental applications include the use of SEM and TEM to characterize fine and ultrafine particulates present in ambient air. In an urban air study [11], Utsunomiya et al. conducted analyses using several TEM techniques to characterize the particulate size associated with heavy metals and to speciate the metals detected. Metals of particular interest for engineered nanomaterials — titanium, iron, and silver — were all detected in nanoparticles. Titanium and iron were present at comparatively high concentrations and were attributable to fractal rock and numerous natural and anthropogenic sources, highlighting the difficulty of determining potential air sources from the manufacture or use of zero-valent iron or titanium dioxide nanoparticles against naturally high backgrounds. Silver was present at low levels, primarily associated with soot particles, and tentatively attributed to background combustion sources.

SEM and TEM provide invaluable information for many purposes. They do, however, have several limitations for environmental applications. Although SEM has a larger field of view than TEM, both SEM and TEM can analyze only a relatively small number of particles at a time. Representativeness for a nonhomogeneous sample is difficult to achieve. The instrumentation is costly and requires a high level of technical expertise to operate properly. The sample preparation and analysis are time-consuming. The particles must be deposited on a support film, and the different ways of achieving this deposition may allow some nanoparticles to aggregate or to fragment, losing some of the characteristics responsible for their activity. For TEM, nonconductive materials must be coated with a conducting material such as graphite, potentially obscuring critical features. On most available instruments, the sample must be at high vacuum during analysis, and results for nanoparticles with volatile components, such as hydrated salts or oxides, may not be representative for the material as it exists outside the vacuum.

Environmental SEM (ESEM) instruments have been developed recently that utilize differential pressure zones. These do allow analyses with the sample at pressures closer to atmospheric, and ESEM instrumentation also can be modified to allow imaging of nanoparticles while in suspension in water or other liquid media. Condensation, evaporation, and transport of water inside carbon nanotubes have been monitored *in situ* with ESEM [11]. Bogner et al. [12] report the analyses of gold and silica nanoparticles and carbon nanotubes dispersed in water using this technique, which they have named "wet scanning transmission electron microscopy," (wet STEM).

5.1.1.2 Scanning Probe Microscopy (SPM)

Scanning probe microscopy (SPM), a relatively newer tool, provides a true three-dimensional surface image. SPM includes a variety of different techniques, including atomic force microscopy (AFM) and scanning tunneling microscopy (STM), which have proven useful for imaging and measuring materials at the nanoscale. SPM techniques are based on a mechanical survey of the surface of an object or particle. A very fine tip mounted on a cantilever scans over the surface of interest, following the surface profile. Interactions between the tip and the surface deflect the tip as it follows the surface profile. The movement of the tip in response to the interaction can be monitored with a laser reflected from the cantilever to a photodiode array (Figure 5.2). STM monitors the weak electrical current induced as the tip is held a set distance from the surface. STM, under some conditions, can provide chemical composition information for the surface. With AFM, the tip responds to mechanical contact forces as well as atom-level interactions between the tip and surface (such as chemical bonding forces, van der Waals forces, or electrostatic forces).

Since their development in the late 1980s, both techniques have found wide application for nanotechnology materials development, as illustrated by the characterization of fullerene particles in Figure 5.3. AFM also holds promise for environmental applications. AFM can be operated at ambient pressure and can characterize a wide range of particle sizes in the same scan, from 1 nm to 8 µm (micrometer). It can analyze particles on a solid substrate at atmospheric pressure or in a liquid medium such as water. It has been used to characterize the morphology and size distribution of nanometer-sized environmental aerosol particles collected from ambient air, as well as for engineered TiO_2 nanoparticles [14]. The size distribution and morphology of natural aquatic colloids, which play important roles in contaminant binding, transport, and bioavailability, also have been characterized with AFM after their absorption onto a mica substrate [15–17]. A detailed discussion of AFM is provided in the review article by Burleson et al. [3]; further information on

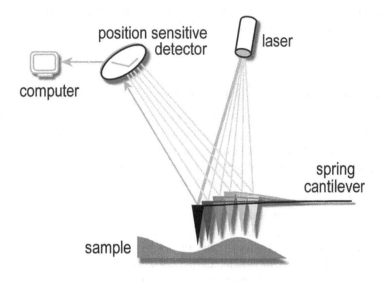

FIGURE 5.2 Atomic force microscopy. (From A. Nadarajah. With permission.)

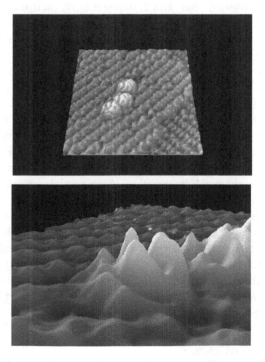

FIGURE 5.3 STM images of buckyballs. (From Nanoscience Instruments. With permission.)

applications of and images from AFM for nanotechnology are available on instrument manufacturers' websites [18, 19].

5.1.2 Compositional Analysis

5.1.2.1 Single Particle Mass Spectrometer

Mass spectrometry forms the basis of several U.S. EPA methods for environmental sample analysis on a bulk basis, providing chemical composition data on an elemental level for metals, and on a molecular level for organics. Mass spectrometry also applies to the analysis of single particles on a real-time basis, although the instrumentation has major differences from mass spectrometers used in U.S. EPA method analyses. The single particle mass spectrometer, first developed in the 1970s for atmospheric aerosol research, analyzes particles from a continuous air stream drawn directly into the ion source. Both organic and inorganic constituents can be detected and identified. The instrument has been widely used for air monitoring studies of particles with aerodynamic diameters in the low micron range [20, 21], but the technology has been extended now to the nanoparticle range.

Most current single particle mass spectrometers are time-of-flight instruments, with some that can detect and analyze particles down to 3 nm in diameter [22]. As a solid particulate or droplet suspended in the air stream enters the source region of the mass spectrometer, a pulsed laser beam desorbs and ionizes the particle components; immediately afterward, a pulsed electric field accelerates all ions of the same charge to the same energy, after which, depending on their mass and charge, they "fly" at different velocities to a charged detector. Both positive and negative ions can be detected in some time-of-flight instruments. These instruments can be field-deployed and have been used in upper atmospheric studies [23] and for on-site ambient air monitoring [24]. Of the nanomaterials specifically discussed in this book, fullerene is the only one for which detection by single particle mass spectrometry has been reported [25].

A recent modification to the technology adds particle size measurement prior to the introduction of the particle into the mass spectrometer source. These instruments, called aerosol time-of-flight mass spectrometers (ATOFMS) [26], employ two distinct time-of-flight technologies. One determines particle size; the other determines particle chemical composition. As a particle enters the instrument, a supersonic expansion of the carrier gas accelerates the particle to terminal velocity. Because smaller particles reach a higher velocity than the larger particles, the aerodynamic diameter can be calculated from the time it takes the particle to travel between two lasers. As the particle passes the second laser and enters the mass spectrometer source, the high-intensity laser of the source is triggered to hit the particle and desorb and ionize particle constituents. These instruments have been used for nanoparticle emission studies from vehicle emissions [27] as well as for atmospheric studies [23].

5.1.2.2   Particle-Induced X-Ray Emission (PIXE)

PIXE measurements can provide major, minor, and trace constituent analyses of nanoparticles. The instrument directs a beam of protons from a high-energy particle accelerator that will knock out core electrons from the atoms of the sample. X-rays are then emitted when outer shell electrons drop into the orbital from which the proton-ejected electron came. The resulting x-ray spectrum of the sample can be used for elemental identifications. The requirement for a particle accelerator to generate the proton beam makes PIXE techniques very costly and available in only a limited number of research laboratories. The technique has been used for trace element analysis of background aerosol particles in the heavily polluted air of Mexico City [28], but it is likely to remain a research tool with limited use.

5.1.3   Surface Area: Product Characterization and Air Monitoring

Surface area is a critical parameter influencing the properties and activity of nanoparticles. In large part, this is believed due to the comparatively high number of atoms on the surface of the particle as opposed to larger particles where most atoms are interior. Surface areas for individual particles can be estimated from the imaging techniques discussed above, but techniques for determining the average surface area for a bulk sample of nanoparticles are more commonly used to monitor production of nanomaterials for specific uses. Some of these methods also are applicable for materials characterization before laboratory exposure studies, and for environmental samples.

5.1.3.1   The Brunauer Emmett Teller (BET) Method

The BET method is named for the three scientists who recognized that particulate surface area can be determined based on the volume of gas that will adsorb to the surface of a given mass of sample. The BET equation relates the volume of gas adsorbed to form a monolayer, the size of the gas molecules, and the mass of the material to derive surface area per unit mass. Commercial analyzers are available that perform this measurement, which may be used during development and production. In a representative research application, BET measurements were relied upon for size characterization of nitrogen-doped titanium dioxide prepared as a photocatalyst for *Escherichia coli* disinfection [29]. Because BET requires a relatively pure bulk sample of a chemically homogeneous material, it has not found application for environmental analyses.

5.1.3.2   Epiphaniometer

The epiphaniometer is a relatively simple device that measures the active surface area of aerosol particles. Particles entering the instrument are charged with radioactive lead ions and then collected on a collection filter. The measured total radioactivity is a measure of the attachment rate, which then allows calculation of the total active surface area of particles in the sample. The requirement for a radioactive source limits the wide use of this instrument, but it has been used in research programs such as mobile laboratory studies of on-road air quality as related to traffic emissions [30].

5.1.3.3   Aerosol Diffusion Charger

The same measurement principle as used for the epiphaniometer is applied in aerosol diffusion chargers but without the requirement for a radioactive source. Ions are produced in a carrier gas by electrical discharge. The ions attach to the surface of the particles, which are then collected in an electrically insulated particle filter. The electric charge is converted to a direct current (DC) voltage signal in an electrometer amplifier. Studies have shown that these devices provide a good estimate of aerosol surface area in ambient air when airborne particles are smaller than 100 nm in diameter [31].

5.1.4   SIZE DISTRIBUTION

Individual particle sizes can be measured accurately with TEM, STEM, and AFM, but those techniques are not time or cost efficient when a complete size distribution is required. Size distribution analyses generally are conducted with aerosols formed when the particles are suspended in air, or when particles are in emulsions or suspensions in a liquid matrix.

5.1.4.1   Electrostatic Classifiers

Electrostatic classifiers operate on the basic principle that the velocity of a charged spherical particle in an electrical field relates directly to its diameter. Particles are suspended in air to form an aerosol, charged, and then introduced into a cylindrical apparatus. The classifier has an outer cylinder that is a ground electrode and an inner rod that can have precisely controlled negative voltage applied. The charged particles are introduced near the wall of the outer cylinder, with a sheath of clean air moving through the cylinder at a constant flow rate. The positively charged particles will move toward the negatively charged center electrode at a rate determined by their operative diameter and the applied voltage. Only those particles within a narrow velocity range will pass through a thin sampling slit near the bottom of the center electrode. Particles exit through this slit into a particle-counting instrument. By scanning the voltage on the central rod, analysts can obtain a full particle-size distribution for the aerosol. It should be noted, however, that the particle size measured is based on the assumption of a spherical shape, and the dimensions of nonspherical particles determined by this technique will correlate with but not necessarily equal those determined by an imaging technique.

Various types of (and names for) electrostatic classifiers are in common use. These include the differential mobility analyzer (DMA), nanodifferential mobility analyzer (NDMA), the differential mobility particle sizer (DMPS), and the scanning mobility particle sizer (SMPS). Electrostatic classifiers can be used in a variety of ways, including real-time monitoring of the length of carbon nanotubes during synthesis [32] or to monitor emissions during use of TiO_2 nanopowder materials [33].

5.1.4.2   Real-Time Inertial Impactor: Cascade Impactors

Cascade impactors have a long history with ambient air monitoring programs, providing size selectivity to the collection of suspended particles. These units take advantage of the differences in settling rates between particles of different aerodynamic diameters. A cascade impactor has co-linear plates in series of pairs through which air is drawn. The first plate of each pair has a small nozzle or nozzles in it to control flow velocity. After the sample passes through the nozzle(s), it is turned sharply before the solid plate, which acts as a collection plate. Particles larger than the stage cut diameter (which is a function of the flow velocity and the distance between the plates) cannot follow the flow stream lines but fall onto the collection plate. Particles smaller than the stage cut diameter continue to the following impactor stages. Ambient air cascades through succeeding stages, which have successively smaller orifices and consequently higher orifice velocities. Collection plates at each successive stage will collect successively smaller particles. While most available units were designed to meet the regulatory requirements to monitor for particulate matter with aerodynamic diameters of 2.5 μm (PM2.5) or less as a category, newer units designed with up to 13 stages can separate particulate down to 30 nm [34]. Samples are time-integrated and may be collected from the plates for further characterization analyses by electron microscopy or other techniques. A micro-orifice uniform deposit impactor (MOUDI) allows collection of nanoparticles in three stages: <32 nm, <18 nm, and <10 nm. These units have been used to characterize nanoparticles from vehicular emissions [35].

5.1.4.3   Electrical Low Pressure Impactor (ELPI)

The electrical low pressure impactor (ELPI) is an extension of cascade impactor technology that includes the multi-stage cascade impactor with detector technology to provide real-time data for both particle size and concentration. This makes it possible to measure rapidly changing conditions in ambient air. The design of the ELPI is based on combining electrical detection principles with low-pressure impactor size classification. The gas sample containing the particles passes through an electrical discharge that ionizes aerosol particles. The charged particles then pass into a low-pressure impactor with electrically isolated collection stages. The electric current carried by the charged particles into each impactor stage is measured in real-time by a sensitive multichannel electrometer. A version designed for ambient or indoor monitoring can detect down to 90 nanoparticles in the 30-nm or smaller range per cubic centimeter (nm/cm³), and can measure a mass as small as 0.005 micrograms per cubic meter (μg/m³) [36]. The ELPI has been used for indoor air monitoring, vehicular emission studies, and ambient air monitoring [37].

5.1.4.4   Dynamic Light Scattering (DLS)

Where the electrostatic classifiers measure the size distributions of particles suspended in air, dynamic light scattering instrumentation determines size distributions for particles suspended in the liquid phase. Light passing through a liquid or suspension of nanoparticles will be scattered, and for nanoparticles, the intensity of the scattered light will fluctuate. This fluctuation results from the random movement of the nanoparticles as a result of their random bombardment by the molecules of the fluid. The velocity and distance of this movement (called Brownian motion), and the subsequent fluctuation of scattered light intensity, depend on the size of the particles because smaller particles are "kicked" further by the solvent molecules and move more rapidly. With a multi-exponential analysis of the scattered light, a particle size distribution can be calculated. The diameter obtained by this technique, called the hydrodynamic diameter, is that of a sphere that would move with the same velocity and to the same distance as the particle being measured. For nonspherical nanoparticles, this diameter will depend on not only the physical dimensions of the particle, but also on its surface structure and on effects from any dissolved material in the sample. The size calculated from DLS measurements is often larger than the dimensions measured by electron microscopy. DLS instrumentation is readily available and relatively straightforward to use, and the technique can be applied in a dynamic fashion to monitor changes in the degree of clustering or agglomeration of nanoparticles *in situ*.

DLS also can be referred to as photon correlation spectroscopy (PLS) or quasi-elastic light scattering (QELS). The newest instrumentation allows measurements down to 1 nm.

DLS is used in studies to predict toxicity or environmental effects, and to confirm the size distribution of material before use and to monitor changes. It has been used to determine the particle size of TiO_2 and fullerene prior to their use in experiments to determine the effect of flow on transport and deposition in porous media [38], and to monitor the aggregation of zero-valent iron particles [39] and TiO_2 [40] in laboratory experiments designed to investigate reasons for the limited mobility of these in environmental settings.

5.2   WORKPLACE AIR MONITORING

The first of five challenges for the safe handling of nanotechnology as identified by scientists in the field [41] is to "develop instruments to assess exposure to engineered nanomaterials in air and water, within the next 3 to 10 years." The exposure of workers to engineered nanoparticles during their production and direct use is of particular concern, and the challenge cites the need for inexpensive personal aerosol samplers capable of measuring and logging the number of nanoparticulates, their surface area, and overall mass concentration in order to assess exposure. As discussed in Chapter 9, nanoparticles can enter the body through respiratory, dermal, and ingestion exposure and then be transported through intercellular pathways. Because the physical characteristics of a nanoparticle (such as size, shape, structure, surface area, and surface activity) determine the body's response, knowing the chemical composition and overall air concentrations solely in terms of any one of these parameters is not enough. Maynard [42] reviews the challenges and technologies for workplace monitoring as was current in 2005.

In some instances, the occupational setting may offer the advantages of limited complexity and available reference material — when the engineered nanoparticles of concern are available in adequate amounts for complete characterization, when there is minimal variability in their physical properties, and when few interferences from other sources in the workplace air are expected. In these instances, the measurement challenge can be separated into two distinct approaches: (1) physical and chemical characterization, which can be completed on the source material by appropriate methods already described; and (2) counting or mass measurements to determine particulate numbers and surface areas for exposure assessment. It should be emphasized, however, that even in relatively controlled environments, the challenges for protective monitoring are considerable. As noted in Chapter 9 of this book, the current state of knowledge on the mechanisms of action and toxicology of specific nanomaterials is very limited. The critical parameters or appropriate range for monitoring for worker safety is not well understood for most nanomaterials; and given the uncertainties, the design of worker safety monitoring programs that are protective and cost-effective remains difficult at best. As presented in Section 5.4, governmental agencies on a global basis have made monitoring for worker safety a research priority.

5.2.1 CONDENSATION PARTICLE COUNTER (CPC)

Condensation particle counters (CPCs) measure the number of particles in an air sample. Commercially available models operate on the principle that small particles serve as condensation nuclei for vapors. A constant flow of air is pulled through the meter, first entering a chamber saturated with water, alcohol, or other organic vapor. The sample and vapor then enter a cooled chamber where the vapor condenses onto the particles, forming droplets large enough to be detected optically. Units currently on the market include hand-held and fixed monitors, with some capable of detecting particles down to 2.5 nm [43]. This technique provides no information on actual particle size, shape, or composition, and particles larger than nanoparticles will be counted unless there is some pre-filtering or separation. This technology is useful for air measurements where the particulates themselves have been characterized by other techniques or for monitoring where the absolute number of particles, either total or below a predetermined size cutoff, will meet the monitoring objective.

5.2.2 SURFACE AREA: TOTAL EXPOSURE

As noted above, particle counting may not be sufficient to evaluate potential risks of exposure to nanoparticles. For each type of nanomaterial, the surface area of individual particles can significantly affect the activity of the material toward biological tissues [44].

Development of a real-time instrument that can monitor exposure as opposed to a single parameter represents an important advance toward ensuring safe working environments for the engineered nanoparticle industry. The electrical aerosol detector (EAD) measures a unique aerosol parameter called aerosol diameter concentration, or total aerosol length. This measurement (reported as mm/cm^3) represents a number concentration multiplied by average diameter, and thus is directly related to surface area. The aerosol diameter concentration, when complemented by CPC data for particle number, can be used to calculate the average particle size. Continuous measurements of aerosol diameter concentration with the EAD correlate well with the surface area of deposited particles and are believed to provide a better estimate of actual inhalation exposure than either the mass or number concentration of particles could. EAD has been used in ambient air studies at the St. Louis Supersite [45].

5.3 SAMPLING AND ANALYSIS OF WATERS AND SOILS FOR NANOPARTICLES

The second subset of the challenge to develop instruments to assess exposure to engineered nanomaterials is "to develop instruments that can track the release, concentration and transformation of engineered nanoparticles in water systems" [41]. Measurements of natural nanoparticles in environmental waters have been reported by a variety of techniques, but there are few reports at this time of field studies designed to detect engineered nanoparticles.

A separation technique called field-flow fractionation (FFF) separates nanoparticles from larger particles, permitting their direct analysis or collection for detailed characterization.

As shown in Figure 5.4, field-flow fractionation is similar to the chromatographic separations typical of environmental analyses for organic contaminants. The water sample passes through a thin flow channel designed so that the flow will be laminar, that is, not turbulent and faster in the center of the column than at the walls. The bottom side of the channel is a membrane that will allow water through but not the particles of interest. A second force is applied perpendicular to the channel flow to generate a cross-flow. All particles in the sample will be pushed downward toward the membranes, but smaller particles will diffuse upward toward the center of the channel to a greater degree and will be in the faster stream lines of the channel flow. The smaller particles in the sample will exit from the channel before the larger particles.

Once separated, natural particles in the nanometer size range can be directly introduced into an inductively coupled mass spectrometer [46] for elemental analysis, collected for ESEM analysis [47], and coupled to a light scattering instrument for particle size measurements [48]. Field-flow fractionation was applied for size distribution analysis of trace concentrations of iron oxi/hydroxide colloids being considered as potential carriers for the radionuclide migration from a nuclear waste repository [49].

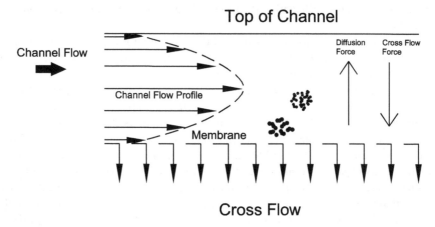

FIGURE 5.4   Flow field fractionation. (From Postnova Analytics, Inc. With permission.)

Field flow fractionation also has been applied to the analysis of nanoparticles in soil and sediment systems. Engineered zinc nanoparticles have been separated from larger soil particles through the preparation of suspensions that are then shaken and allowed to settle gravitationally. The supernatant, containing the less than 1-μm particle fraction, was then separated by field-flow fractionation for further analysis [50].

Analyses of engineered nanoparticles directly in soil or sediment matrices by SEM or TEM imaging techniques is possible when the material has some unique property, such as fluorescence or light absorption, or contains a rare metal or unique organic compound [51]. Nanoparticles of natural origin are ubiquitous, and the detection of engineered particles against background using these techniques, which are at best time-consuming and costly, is not likely to be a practical means of routine environmental assessments.

5.4 NANOTECHNOLOGY MEASUREMENT RESEARCH AND FUTURE DIRECTIONS

Both within the United States and internationally, private and governmental organizations have recognized the need for improved analytical tools for nanotechnology. The American Society for Testing and Materials (ASTM) Committee E56 on Nanotechnology was formed in 2005 to develop standards and guidance for nanotechnology and nanomaterials. The first ASTM standard, published in July 2007, precisely defines the language for nanotechnology [52]. This should allow more consistent and effective technical communication within the diverse fields involved in nanotechnology and with the public.

In late 2005, the International Organization for Standardization (ISO) established a new technical committee, ISO T/C229 Nanotechnologies, with three working groups. The United States is represented on the Measurement and Characterization Workgroup, WG2. The Draft Business Plan for T/C229 [53] details the high-priority needs and strategies for this group. A limited number of standards relating to nanoparticle measurements have been published, with several more in progress. ISO/TR 27628:2007 contains guidelines on characterizing occupational nanoaerosol exposures, with a discussion of applicable measurement terms. Specific information is provided on methods for bulk aerosol characterization and single-particle analysis. Other standards currently near completion include N 270 TS: Terminology and definitions for carbon nanomaterials; N 271 TS: Format for reporting the engineered nanomaterials content of products; and N 272 TR: Guide to nanoparticles measurement methods and their limitations.

5.4.1 United States

5.4.1.1 NIOSH

In the United States, the National Institute for Occupational Safety and Health (NIOSH) established the Nanotechnology Research Center (NTRC) in 2004 to coordinate and facilitate research on the impact of nanotechnology in the workplace. The NTRC recognized that in order to evaluate risks, accurate measurement data would

TABLE 5.2
NIOSH Goals for Nanoparticle Measurement

1. Evaluating methods of measuring mass of respirable particles in the air and determining if this measurement can be used to measure nanomaterials
2. Developing and field-testing practical methods to accurately measure airborne nanomaterials in the workplace
3. Developing testing and evaluation systems to compare and validate sampling instruments

TABLE 5.3
NIOSH Research Agenda

Fiscal Year	NIOSH Nanoparticle Research Strategic Plan and Timeline (Measurement and Analysis Programs)
2005	Surveillance Phase I: Identify and gather baseline information. Develop techniques for online surface area measurement.
2006	Conduct measurement studies of nanoparticles in the workplace. Analyses of filter efficiency for nanomaterials.
2007	Evaluate surface area-mass metric results. Establish a suite of instruments and protocols for nanomaterial measurements. Conduct measurement studies of nanoparticles in the workplace. Further development of online and offline nanoparticle measurement methods.
2008–2009	Develop performance results for nanoparticle measurement instruments and methods. Complete evaluation of viable and practical workplace sampling devices and methods for nanoparticles (affordable, portable, effective). Quantification of systemic nanoparticle concentrations in laboratory animals after pulmonary exposure to nanospheres and nanofibers.

be needed. Measurement method development objectives, as listed in Table 5.2, were accordingly included as critical topics for their strategic workplace goals.

The NTRC established several partnerships with other agencies, including U.S. EPA, NIST, the Department of Defense (DOD) and the Department of Energy (DOE), and ASTM. In addition, the NTRC is partnering with and in some instances supporting research at academic institutions, instrument manufacturers, and private industry. Table 5.3 summarizes method analysis studies included in the research program planned for the period 2005 through 2009 [54].

Accomplishments and publications for 17 completed and ongoing research programs are listed in the 2007 NIOSH report entitled "Progress toward Safe Nanotechnology in the Workplace" [55]. "Project 1, Generation and Characterization of Occupationally Relevant Airborne Nanoparticles," includes numerous accomplishments relevant to the characterization and workplace measurement of carbon nanotubes and TiO_2. Project 11, Nanoparticles in the Workplace, is designed to develop partnerships with industry, academia, and other government agencies for research and development of monitoring instrumentation and protocols. Project 13, The Measurement and Control of Workplace Nanoparticles, will provide a basis for

TABLE 5.4

U.S. Government Supported Research on Nanotechnology Environmental Measurements

Project Title	Sponsor	Anticipated End Year
Biological Fate and Electron Microscopy Detection of Nanoparticles during Wastewater Treatment	EPA	2010
Development of Detection Techniques and Diagnostics for Airborne Carbon Nanotubes	DOE	2007
Fate and Transport of Carbon Nanotubes in Unsaturated and Saturated Soils	EPA	2008
Identifying and Regulating Environmental Impacts of Nanomaterials	NSF	2007
Monitoring and Characterizing Airborne Carbon Nanotube Particles	NIOSH	2008
New Instruments for Real-Time, High Resolution Characterization of Nanoparticles in the Environment	NSF	2007

understanding how nanoparticles are released in the workplace and how they can be monitored and exposure controlled. Research Project 14 is an exposure study of TiO_2 in manufacturing and end-user facilities using a variety of monitoring techniques.

5.4.1.2 U.S. Government-Sponsored Research

The Project on Emerging Nanotechnologies of the Woodrow Wilson International Center for Scholars in Arlington, Virginia, is developing an inventory of government-sponsored research into the environmental, health, and safety implications of nanotechnology [56]. While the intent of this inventory is to include research projects on an international basis, the current listing is dominated by projects supported by U.S. agencies. Included in these are several funded by the U.S. EPA, the Department of Energy (DOE), the National Science Foundation (NSF), and NIOSH that should provide information on environmental measurements. Current projects included in this inventory that are of particular relevance to environmental measurements include those listed in Table 5.4.

5.4.1.3 National Institute of Standards and Technology (NIST)

In early 2006, the National Institute of Standards and Technology (NIST) launched a new state-of-the-art Center for Nanoscale Science and Technology (CNST) [57]. The CNST is specifically dedicated to developing the measurement methods and tools needed to support all phases of the nanotechnology industry. While the Center's focus is not specifically toward environmental methods or analyses, measurement advances for discovery, development, and manufacturing of nanoparticles should have applicability in monitoring their presence and effects in the environment.

NIST has responsibility for the development and supply of standard reference materials for analyses of various chemicals and materials. Academic laboratories as well as commercial facilities use these standard reference materials to verify the accuracy of their data. NIST currently provides certified polystyrene spheres for nanoparticle size analyses. NIST [57] also reports a recent development of a prototype atomic "ruler" for calibrating dimensional measurements of 100 nm and below. This ruler will be able to document the accuracy of scanning electron or atomic force microscopy data. NIST now is transferring the technology to a commercial standards supplier.

5.4.2 EUROPEAN UNION

The European Union (EU) launched its largest ever funding program for research and technological development on January 1, 2007. The EU Member States have earmarked a total of €3.5 billion (approx. U.S. $4.5 billion) for funding nanotechnology related research over the period 2007 to 2013. The program calls for proposals for a wide range of activities related to the risk assessment of nanomaterials [58]. First among five areas where proposals are invited is NMP-2007-1.3-1, "Specific, easy to use portable devices for measurement and analysis." Based on the belief that workplace exposure is the area of greatest concern, the objective of this work will be "to develop and validate affordable, portable, adequate sampling and measurement equipment for monitoring working environments (i.e., quantification and characterization of airborne nanoparticles in particular)."

German governmental agencies, including the Federal Institute for Occupational Safety and Health (BAuA), the Federal Institute for Risk Assessment (BfR), and the Federal Environment Agency (UBA), have jointly developed a nanotechnology research strategy and are currently conducting a limited number of projects [59]. Two of these are designed to test instrumentation or modify and validate existing measurement methods to be applicable for workplace measurements of nanoparticles.

5.4.3 ASIA-PACIFIC

The Industrial Technology Research Institute (ITRI) of Chinese Taipei is undertaking an international project awarded by the Asia-Pacific Economic Cooperation (APEC) Industrial Science and Technology Working Group (IST WG) to assist in the establishment of the Technological Cooperative Framework on Nanoscale Analytical and Measurement Methods among APEC economies. This cooperative project was formed to create an avenue for sharing advances in nanometer analytical measurement methods and to promote the best available technology to meet the needs for nanoscale standards [60]. The United States is one of six nations participating in this project.

As part of the project, the NanoTechnology Research Center (also using the acronym NTRC) of ITRI is organizing an interlaboratory comparison study on nanoparticle characterization. The aim of the comparison is to establish the effectiveness and comparability of different measurement methods across different laboratories on nanometer-scale particles. This multi-year program will include a series of measurement challenges using standardized material, as shown in Table 5.5.

The results for 2005 interlaboratory comparison measurements of size and diameter have been published [60]. The NTRC provided samples of polystyrene

TABLE 5.5
APEC Nanometer Measurement Interlaboratory
Comparison Studies for Nanoparticles

Year	Characteristic
2005	Size/diameter
2006	Surface area, refractive index, light absorption/reflection
2007	Number and distribution, conductivity, dispersion

spheres with diameters of 30, 50, and 100 nm to participating laboratories. The study generated a total of 32 data sets from 15 participating laboratories. Particle size measurements were made using DLS, SEM, TEM, and SPM instrumentation. The results for the 30-nm particles were satisfactory from all measurements, while two measurements for the 50-nm particles and four for the 100-nm sample fell outside three standard deviations of the mean. Measurements taken by TEM for the 30-nm and 100-nm particles were significantly below the expected diameters and below the results from the other techniques. Samples were distributed for the 2006 studies, and 16 laboratories have reported results, but these have not been made publicly available at the time of writing.

5.5 SUMMARY

Reliable and accurate measurements of nanoparticle physical and chemical properties have been recognized as critical elements required for meaningful assessments of impact and risk. While numerous technologies do exist, many challenges to measuring engineered nanoparticles in the environment have yet to be addressed. The combination of their small size and the range of attributes that may factor into their activity requires a complex matrix of complementary analyses and methods, for many critical parameters have yet to be devised for nanoparticles in environmental settings. Environmental analyses of nanoparticles are far from routine or readily available at this point, but the increased interest and focus of governmental agencies and research organizations allows for optimism.

REFERENCES

1. Powers, K.W., M. Palazuelos, B. Moudgil, and S. Roberts. 2007. Characterization of the size, shape, and state of dispersion of nanoparticles for toxicological studies. *Nanotoxicology,* 1:42–51.
2. U.S. Environmental Protection Agency. 2007. *EPA Nanotechnology White Paper.* U.S. EPA 100/B-07/001.
3. Burleson, D.J., M.D. Driessen, and R.L. Penn. 2004. On the characterization of environmental nanoparticles. *J. Env. Science Health,* 39:2707–2753.
4. Durbin, T.D., J.M. Norbeck, D.R. Cocker, and T. Younglove. 2004. Particulate Matter Mass Measurement and Physical Characterization — Techniques and Instrumentation for Laboratory Source Testing. Final Report ARB Literature Searches.

5. Scalera, J. 2006. Detection and Characterization of Nanomaterials in the Environment. Presented at Session 3 of *the EPA Symposium on Nanotechnology and the Environment.* http://www.epa.gov/swerrims/nanotechnology/events/OSWER2006/sessionl. html.

6. Mansfield, J.F. 2000. An introduction to the transmission electron microscope and what it can do. http://www.emal.engin.umich.edu/courses/TEMChem2000/index.htm.

7. Grassian, V., P.T. O'Shaughnessy, A. Adamcakova-Dodd, J.M. Pettibone, and P.S. Thorne. 2007. Inhalation exposure study of titanium dioxide nanoparticles with a primary particle size of 2 to 5 nm. *Environ Health Perspect.,* 115:397–402.

8. Lovern, S. and R. Klaper. 2006. *Daphnia Magna* mortality when exposed to titanium dioxide and fullerene (C60) nanoparticles. *Environ. Tox. Chem.,* 25:1132–1137.

9. Rothen-Rutishauser, B.M., S. Schurch, B. Haenni, N. Kapp, and P. Gehr. 2006. Interaction of fine particles and nanoparticles with red blood cells visualized with advanced microscopic techniques. *Environ. Sci. Technol.,* 40:4353–4359.

10. Sipzner, L., J. Stettin, Z. Pan, et al. 2006. The penetration of titanium dioxide nanoparticles: from dermal fibroblasts to skin tissue. American Physics Association. http://meetings.aps.org/link/BAPS.2006.MAR.V26.2.

11. Utsunomiya, S., K. Jensen, G. Keeler, and R. Ewing. 2004. Direct identification of trace metals in fine and ultrafine particles in the Detroit urban atmosphere. *Environ. Sci. Technol.,* 38:2289–2297.

12. Rossi, M.P., H. Ye, Y. Gogotsi, S. Babu, P. Ndungu, and J.-C. Bradley. 2004. Environmental scanning electron microscopy study of water in carbon nanopipes. *Nano Lett.,* 4:989–993.

13. Bogner, A., G. Thollet, D. Bassett, P.-H. Jouneau, and C. Gauthier. 2005. Wet STEM: a new development in environmental SEM for imaging nano-objects included in a liquid phase. *Ultramicroscopy,* 104:290–301.

14. Köllensperger, G., G. Friedbacher, A. Krammer, and M. Grasserbauer. 1999. Application of atomic force microscopy to particle sizing. *Fres. J. Anal. Chem.,* 363:323–332.

15. Baalousha, M. and J.R. Lead. 2007. Characterization of natural aquatic colloids (<5 nm) by flow field fractionation and atomic force microscopy. *Environ. Sci. Technol.,* 41:1111–1117.

16. Doucet, F., L. Maguire, and J. Lead. 2004. Size fractionation of aquatic colloids and particles by cross-flow filtration: analysis by scanning electron and atomic force microscopy. *Anal.Chim. Acta,* 522:59–71.

17. Lead, J.R., D. Muirhead, and C.T. Gibson. 2005. Characterization of freshwater natural aquatic colloids by atomic force microscopy (AFM). *Environ. Sci. Technol.,* 39:6930–6936.

18. Pacific Nanotechnology. 2007. Atomic Force Microscopes – Tutorial Page. http://www.pacificnano.com/afm-tutorial.html.

19. Nanoscience Instruments. 2007. Nanoscience Education. http://www.nanoscience.com/education/AFM.html.

20. Gard, E., J.E. Mayer, B.D. Morrical, T. Dienes, D.P. Fergenson, and K.A. Prather. 1997. Real-time analysis of individual atmospheric aerosol particles: design and performance of a portable ATOFMS. *Anal. Chem.,* 69:4083–4091.

21. Silva, P. and K. Prather. 2000. Interpretation of mass spectra from organic compounds in aerosol time-of-flight mass spectrometry. American Physical Society March 2006 Meeting. *Anal. Chem.,* 72:3553–3562.

22. Wang, S., C.A. Zordan, and M.V. Johnston. 2006. Chemical characterization of individual, airborne sub-10-nm particles and molecules. *Anal. Chem.,* 78:1750–1754.

23. Sullivan, R.C. and K.A. Prather. 2005. Recent advances in our understanding of atmospheric chemistry and climate made possible by on-line aerosol analysis instrumentation. *Anal. Chem.,* 77:3861–3886.

24. Kolb, C.E., S.C. Herndon, J.B. McManus, et al. 2004. Mobile laboratory with rapid response instruments for real-time measurements of urban and regional trace gas and particulate distributions and emission source characteristics. *Environ. Sci. Technol.*, 38:5694–5703.

25. Reilly, P.T.A., R.A. Gieray, W.B. Whitten, and J.M. Ramsey. 2000. Fullerene evolution in flame-generated soot. *J. Am. Chem. Soc.*,122:11596–11601.

26. TSI. 2004. Series 3800 Aerosol Time-of-Flight Mass Spectrometers with Aerodynamic Focusing Lens Technology. http://www.tsi.com/documents/3800.pdf.

27. Sodeman, D.A., S.M. Toner, and K.A. Prather. 2005. Determination of single particle mass spectral signatures from light-duty vehicle emissions. *Environ. Sci. Technol.*, 39:4569–4580.

28. Chianelli, R.R., M.J. Yácaman, J. Arenas, and F. Aldape. 1998. Atmospheric nanoparticles in photocatalytic and thermal production of atmospheric pollutants. *J. Haz. Substances Res.*, 1:1–17.

29. Li, Q., R. Xie, Y.W. Li, E.A. Mintz, and J.C. Shang. 2007. Enhanced visible light induced photocatalytic disinfection of *E. coli* by nitrogen doped titanium oxide. *Environ. Sci. Technol.*, 41:5050–5056.

30. Kittelson, D.B., W.F. Watts, and J.P. Johnson. 2001. Fine Particle (Nanoparticles) Emissions on Minnesota Highways. Minnesota Department of Transportation.

31. Nziadchristos, L., A. Polidori, H. Phuleria, M. Gillen, and C. Sioutasm. 2007. Application of a diffusion charger for the measurement of particle surface concentration in different environments. *Aerosol Sci. Technol.*, 41:511–580.

32. Kim, S.H. and M.R. Zachariah. 2005. In-flight size classification of carbon nanotubes by gas phase electrophoresis. *Nanotechnology,* 16:2149–2152.

33. Hsu, L.-Y. and H.-M. Chein. 2007. Evaluation of nanoparticles emissions for TiO_2 nanopowder coating material. In *Nanoparticles and occupational health*, Eds. A.D. Maynard and D.Y.H. Pui, p. 157–163. Springer Netherlands.

34. Dekati. 2007. Specification Sheet: Dekati Low Pressure Impactor. http://www.dekati.com/cms/dlpi/specifications.

35. Wu, Y.-S., G.C. Fang, S.-Y.Chang, J.-Y. Rau, S.-H. Huang, and C.-K.Lin. 2006. Characteristic study of ionic species in nano, ultrafine, fine and coarse particle size mode at a traffic sampling site. Toxicol. Indust. Health, 22: 27–37.

36. Dekati. 2007. Outdoor Air Electrical Low Pressure Impactor. http://www.dekati.com/cms/outdoor_air_elpi.

37. Kittelson, D.B., W.F. Watts, J.P. Johnson, and M.K. Drayton. 2001. Fine Particle (Nanoparticle) Emissions on Minnesota Highways. Presented at the 7th Diesel Engine Emissions Reduction (DEER) Workshop. http://www.me.umn.edu/centers/mel/reports/DEER2001.pdf.

38. Lecoanet, H., J. Rottero, M. Weisner. 2004. Laboratory assessment of the mobility of nanomaterials in porous media. *Environ. Sci. Technol.*, 38:5764–5768.

39. Phenrat, T., N. Saleh, K. Sirk, R. Tilton, and G. Lowry. 2007. Aggregation and sedimentation of aqueous nanoscale zerovalent iron dispersions. *Environ. Sci. Technol.*, 41:284–290.

40. Dunphy Guzman, K., M. Finnegan, and J. Banfield. 2006. Influence of surface potential on aggregation and transport of titania nanoparticles. *Environ. Sci. Technol.*, 40:7688–7693.

41. Maynard, A. 2006. Safe handling of nanotechnology. *Nature* 444:267-269.

42. Maynard, A. 2005. Engineered Nanomaterials: Measurement in the Occupational Setting. Project on Emerging Technologies. Presented at *ECETOC Nanomaterials*, Barcelona, Spain.

43. BPA Air Quality Solutions, Inc. 2007. Nano Particle Counters Product Reviews. http://www.particlecounters.org/nano/.

44. Oberdörster, G., E. Oberdörster, and J. Oberdörster. 2005. Nanotoxicology: an emerging discipline evolving from studies of ultrafine particles. *Environ. Health Perspect.,* 113:823–839.
45. Han, H.S., S. Kaufman, J. Turner, W. Wilson, and D.Y. Pui. 2005. Electrical aerosol detector (EAD) measurements at the St. Louis Supersite. Presented at *AAAR PM Supersites Program and Related Studies International Specialty Conference,* Atlanta, GA, 7–11 February. http://oaspub.epa.gov/eims/eimsapi.dispdetail?deid=95475#top.
46. Stolpe, B., M. Hassellov, K. Andersson, and D. Turner. 2005. High resolution ICPMS as an on-line detector for flow field-flow fractionation: multi-element determination of colloidal size distributions in a natural water sample. *Anal. Chim. Acta,* 535:109–121.
47. De Momi, A. and J. Lead. 2006. Size fractionation and characterization of fresh water colloids and particles: split-flow thin cell and electron microscopy analyses. *Environ. Sci. Technol.,* 40:6738–6743.
48. Baalousha, M., F.V.D. Kammer, M. Motelica-Heino, H.S. Hidal, and P. Le Coustumer. 2006. Size fractionation and characterization of natural colloids by flow-field flow fractionation coupled to multi-angle laser light scattering. *J. Chromatogr,* 1104:272–281.
49. Bouby, M., H. Geckeis, T.N. Manh, et al. 2004. Laser-induced breakdown detection combined with asymmetrical flow field-flow fractionation: application to iron oxi/hydroxide colloid characterization. *J. Chromatogr,* 1040:97–104.
50. Gimbert, L.J., R.E. Hamon, P.S. Casey, and P.J. Worsfold. 2007. Partitioning and stability of engineered ZnO nanoparticles in soil suspensions using flow field-flow fractionation. Environ. Chem., 4:8–10.
51. von der Kammer, F., T. Hoffman, and M. Hasselhov. 2006. Nanopollution: How to Gain Knowledge on the Behavior and pthwaysof Engineered Nanoparticles in the Aquatic Environment. Poster presentation, *SETAC 2006.* www.univie.ac.at/env-geo/Publications/Poster/vdKammer_SETAC_2006_NANOPOLLUTION.pdf
52. ASTM. 2007. Standard E 2456-06, Terminology for Nanotechnology.
53. ISO. 2007. Business Plan ISO/TC 229 Nanotechnologies (Draft) ISO/TC 229 N230.
54. NIOSH. 2005. Strategic Plan for NIOSH Nanotechnology Research: Filling the Knowledge Gap. Draft. Nanotechnology Research Program.
55. NIOSH. 2007. Progress toward Safe Nanotechnology in the Workplace. DHHS (NIOSH) Publication No. 2007–123.
56. Woodrow Wilson International Center for Scholars. Nanotechnology Health and Environmental Effects: An Inventory of Government-Supported Research. http://nanotechproject.org/index.php?id=29.
57. NIST. 2006. Fact Sheet from NIST: Introducing the NIST Center for Nanoscale Science and Technology. http://www.nist.gov/public_affairs/factsheet/CNST_factsheet.htm.
58. European Commission. 2005. Seventh Framework Programme of the European Community for Research and Technological Development including Demonstration Activities (FP7).
59. BAUA, BfR, UBA. 2006. Nanotechnology: Health and Environmental Risks of Nanoparticles — Research Strategy (Draft).
60. Fu, W.-E. 2006. APEC Project: Technological Cooperative Framework on Nanoscale Analytical and Measurement Methods: Project Overview. Presented at the *2006 APEC Nanoscale Technology Forum.*

6 Environmental Fate and Transport

Chris E. Mackay and Kim M. Henry
AMEC Earth & Environmental

CONTENTS

The movement and transformation of materials within an environmental setting is a very important consideration when evaluating the risks associated with their release. The greater a material's stability, in terms of low chemical reactivity and ready suspension in fluid environmental media, the greater its potential for distribution and therefore the wider the potential scope of exposure (area, number of receptors, types of habitats, etc.).

6.1 INTRODUCTION

The environmental fate and transport of a given chemical can usually be characterized or predicted based on a relatively small set of characteristics. These typically include phase properties (boiling point, melting point, vapor pressure); affinity properties (air/water, water/soil, etc.); media reactivity (hydrolysis, oxidoreduction, photoreactivity); and biological degradation rates. Most models of environmental fate and transport use a combination of some or all of these properties to predict concentrations within various environmental media. The potential for environmental risk can then be determined from these predicted concentrations based on the toxicity of the materials.

This chapter examines the fate and transport of free nanomaterials in the environment. In some cases, nanomaterials may be considered in a manner identical to smaller molecular materials. Other cases require special methods to account for differences in the physical and chemical properties of nanomaterials as well as their peculiar phase properties. (See Chapter 2 for a discussion of the critical properties of nanomaterials.)

Figure 6.1 illustrates the primary forces that determine the fate and transport of nanoparticles in suspension. Upon an initial release of disperse nanoparticles, buoyancy suspends the nanoparticles in the fluid. Van der Waals forces, relatively weak forces resulting from transient shifts in electron density, cause the nanoparticles to

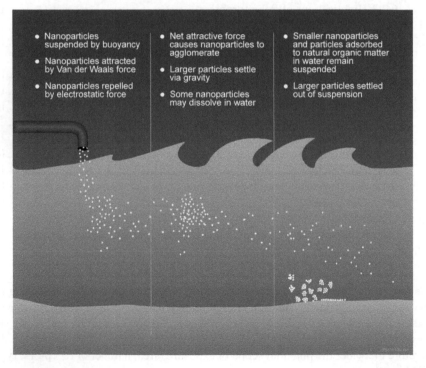

FIGURE 6.1 Conceptual model of primary forces determining fate and transport of nanoparticles in solution.

be attracted to one another and to other environmental constituents. (The term "physisorption" refers to adsorption as a result of van der Waals forces.) Nanoparticles will tend to agglomerate unless this physisorption is inhibited. As the size of the agglomerates increases, buoyancy is reduced and the force of gravity causes the particles to settle out of suspension. If the nanoparticles have similar electrostatic surface charges, however, the repulsive force will counter the attraction resulting from van der Waals forces and keep particles in suspension. Nanoparticles also can adsorb to natural organic matter. That may either increase the particles' buoyancy or disrupt subsequent agglomeration, thereby allowing the nanoparticles to remain suspended. Other environmental interactions such as dissolution or biodegradation also can reduce the concentration of nanoparticles in suspension. As a result of the various forces acting on nanoparticles, which become even more complex than this simple conceptual model when considering transport through soil, the concentration of nanoparticles in solution does not remain at equilibrium but changes over time and over distance from the discharge point.

Sections 6.2 and 6.3 describe the forces that affect the fate and transport of nanoparticles. (Section 6.6 lists the symbols used in mathematical equations in those sections.) As with any model, the mathematics can approximate only real-world complexities. The nanoparticles' characteristics such as a shape or variance in composition will affect the material's chemical properties. Further, the environmental characteristics of the suspending medium such as the pH, hardness, mineral content, ionic strength, types and amounts of dissolved organic matter, and especially the characteristics of sediment/soil will affect the environmental fate and transport of nanomaterials. Section 6.4 summarizes research findings regarding the fate and transport of the target nanomaterials, which account for the effects of some of those characteristics.

6.2   NATURE OF NANOMATERIALS IN THE ENVIRONMENT

Special considerations unique to predicting the fate and transport of nanomaterials can be divided into two general groups: (1) those related to the physical manifestation of the materials, and (2) those related to special chemical properties that affect their reactivity and interactions with their surroundings. Each is discussed below.

6.2.1   PHYSICAL MANIFESTATION OF NANOMATERIALS: PARTICLE SIZE DISTRIBUTION AND FORMATION OF MOBILE SUSPENSIONS

Nanoparticles can form suspensions in air or water, and can be transported through the environment in such suspensions. The suspension of nanoparticles is not an equilibrium phenomenon, but depends in part on the particle size and changes in particle size that result from collisions and reactions in the environment, as discussed below. Other factors that affect the suspension of nanoparticles are discussed in subsequent sections.

With few exceptions, preparations of nanomaterials are not of uniform particle size. Rather, nanopreparations consist of a distribution of varying particle sizes. When a nanomaterial is released into a fluid environment, such as air or water, the size distribution will begin immediately to change as the result of differential settling based on the particle size. This results from the vector settling force ($F_\downarrow$), which is a function of buoyancy and gravity (g).

$$F_\downarrow = \rho_x V_x g \quad Gravity$$

$$F_\downarrow = -(\rho_f V_x g) \quad Buoyancy \tag{6.1}$$

$$\therefore F_\downarrow = V_x g (\rho_x - \rho_f) \quad Settling\ Force$$

When expressed as force vectors, it becomes clear that the smaller the nanoparticle's volume (V_x), the lower the force vector, regardless of the difference in either particulate (ρ_x) or fluid (ρ_f) densities. The extremely small particle size of nanomaterials results in a very low settling force due to the small magnitude of V_x. In short, over time, the concentration of suspended nanoparticles will decline as the larger particles settle out of suspension while the smaller particles remain in suspension.

The rate at which particles settle out of suspension determines the potential for transport through the environment and the ease of removal through air or water treatment processes. The settling or terminal velocity (v_x) is a function of the settling force and the fluid's resistance to passage or viscosity (η) as follows:

$$v_x = \frac{2}{9} \cdot \frac{r^2 g}{\eta} \cdot (\rho_x - \rho_f) \tag{6.2}$$

where r is the effective particle radius. Table 6.1 provides examples of the effect of particle radius on the settling rate of titanium dioxide in air and water. These examples show that as the particle size decreases, the rate of settling decreases substantially and thus the particles can stay in suspension more readily.

At particle sizes below 100 nm, the settling velocity has a magnitude akin to rates of Brownian motion, which is the random movement of small particles suspended in a fluid resulting from the thermal velocity of the particles in the suspending medium. As a result, the particles can form a stable suspension. Such systems, referred to as sols, can occur in fluids such as water (hydrosol) or gases such as atmospheric air (aerosol).

Suspensions of nanoparticles may not be true solutions. This is because the suspension is not the result of an equilibrium condition, but rather the result of very

TABLE 6.1
Sedimentation Rate for TiO$_2$ Spheres of Varying Size in Water and Air

	(cm/hr)	
Particle Diameter	Settling Rate in Water (v_x)	Settling Rate in Air (v_x)
1 mm	7×10^2	3×10^4
1 μm	7×10^{-4}	3×10^{-2}
100 nm	7×10^{-6}	3×10^{-4}
10 nm	7×10^{-8}	3×10^{-6}

Note: Pressure = 1 atm; Temperature = 25°C.

slow settling kinetics. As a result, nanoparticles can be said to possess an apparent solubility (k_{as}) that can be described in a manner similar to that for a solution as follows:

$$k_{as} = \frac{[X]_f}{[X]_s} \qquad (6.3)$$

where $[X]_f$ represents the concentration of nanoparticle X in sol and $[X]_s$ represents the concentration in the solid, non-sol form. If it is assumed that the material is initially introduced into the fluid medium in the nanoparticulate form, the settling rates are within a range of thermal kinetics, and hence absolute temperature (T) becomes a factor in determining the equilibrium concentration of the particles in the sol. An expression for k_{as} can be derived using the Boltzmann equation as follows:

$$\ln k_{as} = \ln \frac{[X]_f}{[X]_s} = -\int \frac{V_x g(\rho_x - \rho_f)}{kT} \cdot h \, dh \qquad (6.4)$$

where k is the Boltzmann constant, T is absolute temperature, and h is the linear measure of particle separation. At saturation, the amount in non-suspension (i.e., $[X]_s$) will have no real effect on the amount in suspension, Hence the equilibrium equation can be expressed solely based on the aqueous concentration of the nanoparticle as follows:

$$\ln k_{as} = -\int \frac{V_x g(\rho_x - \rho_f)}{kT} \cdot h \, dh \qquad (6.5)$$

The integration of the Boltzmann equation allows a first approximation of the total suspended nanoparticulate concentration at equilibrium as follows:

$$\ln(k_{as}) = \int_{x=0m}^{x=0.01m} -\frac{V_x g(\rho_x - \rho_{aq})}{kT} \cdot h \, dh$$

$$= -\frac{V_x g(\rho_x - \rho_{aq})}{2kT} \cdot 0.01^2 \qquad (6.6)$$

Therefore:

$$[X]_{aq} = e^{-\frac{V_x g(\rho_x - \rho_{aq}) \cdot 0.01^2}{2kT}}$$

This derivation shows that the particulate concentration and temporal stability of heterogeneous sols depend on the size of the particles. If the nanoparticles' size is stable, then the suspension will be stable (excluding disruption by outside forces). Thus, nanoparticles can form metastable suspensions. However, if the particles agglomerate with like particles or other constituents in air or water, then the suspension will not be stable. This phenomenon is discussed further in Section 6.2.2.

This method provides a means to predict the concentration of nanomaterials in a hydrosol or aerosol based on the physical properties of the materials and the interplay of particle size and density (Figure 6.2). For materials with a density less

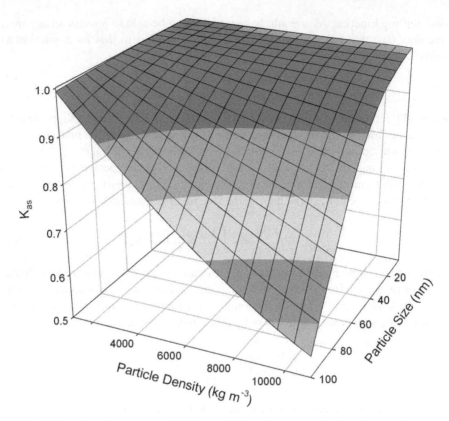

FIGURE 6.2   Plot of apparent solubility coefficient (k_{as}) against particle size and density.

than that of lead, (11.5 g/cm³), all particles within the definition of a nanomaterial
will possess high k_{as} values and capacity for metastable suspension (Figure 6.3). This
method can be applied to materials containing particles in a range of sizes by defining the volume as a distribution function ($f(V_x)$). Figure 6.4 provides an example of
this type of application to an aqueous suspension of nanoparticle-sized zero-valent
iron (nZVI).

     As noted above, the derivation of this method assumed that the nanomaterials
are inert and do not interact with environmental constituents. If not, then the integration of the Boltzmann model represents only the initial situation. To determine the
stability of nanoparticle suspensions in reactive environments, dynamic time-course
chemical reactions must be taken into account to predict the nanomaterial's sol stability and thereby its potential for transport and receptor exposure.

6.2.2   CHEMICAL FORCES ACTING ON NANOMATERIALS

If nanoparticle size changes as the result of interactions within the environment,
then the kinetics of the suspension will change. For example, agglomeration resulting from the chemical interactions of the nanoparticles with like particles or with
certain environmental constituents may increase the effective particle size. When
this increase in size reduces the particles' buoyancy sufficiently, they no longer stay

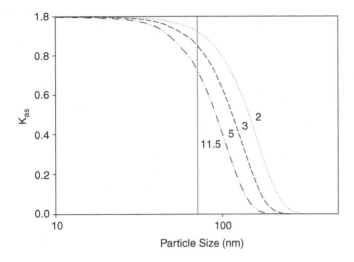

FIGURE 6.3 Calculated apparent solubility for particles of various size and densities; numbers represent particle densities in g/cm³.

in suspension. (Conversely, and as illustrated in Section 6.4, adsorption to dissolved organic matter in surface water can keep some nanoparticles in suspension.)

Within the environment, changes in particle size usually occur as the result of three types of processes: (1) solution/dissolution, (2) adsorption, and (3) agglomeration. Because nanomaterials are defined by initial particle size and not by composition, it is difficult to generalize and predict their chemical properties. However, a few assumptions can be made based on common requirements necessary to form stable nanoparticles:

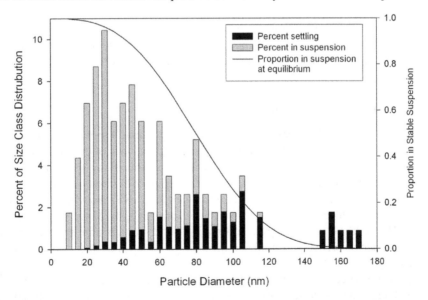

FIGURE 6.4 Projected proportional suspension of zero-valent iron nanoparticles ($\rho_x = 7000$ kg/m³) in aqueous suspension based on the distribution of Nurmi et al. [1].

1. Nanomaterials must be internally structured, based on stable covalent bonds, and will not be immediately soluble in environmental fluid media.
2. The chemical activity of the particle is based on its surface chemistry, which is a function of both its composition and its structure.
3. The nanomaterials will tend not to have either strong nucleophilic or electrophilic affinities; otherwise they would not be stable in particulate form. Therefore, in the absence of harsh agents, they will tend to interact with the environment via weaker ionic and van der Waals interactions.

Predicting the surface behavior of nanomaterials can be very difficult because the architecture of the particle can dramatically affect both energy transfer and electron distribution. This can be particularly true for heterogeneous particles where partial charge sharing or excitation quenching can occur. However, if it is assumed that the initial nanoparticle is indivisible, then the potential for environmental interactions is limited to the interactions of the surface layer. Therefore, by characterizing the surface chemistry, it would be possible to determine the types of interactions that are likely to occur in natural air or water environments. These interactions would determine the most likely physical/chemical fate, and thereby the ultimate disposition of the material once released.

Surface chemistry interactions can be defined using a specific generalized force field summation for colloidal systems developed by Derjaguin, Landau, Verwey, and Overbeet (DLVO) [2]. In the DLVO summation, the total force field (F_T) includes van der Waals forces (F_{vdw}), the forces of solvency (F_s), and electrostatic repulsive forces (F_R) as follows:

$$F_T = F_R + F_{vdw} + F_s \qquad (6.7)$$

These forces, while typically weak, become the significant driving forces for nanomaterials because of the particles' high Brownian velocity and low inherent inertia. Each of these forces, and their implications for the transport of nanoparticles, is discussed below.

6.2.2.1 Electrostatic or Coulomb Force

The electrostatic repulsive or Coulomb force (F_R) represents a specific point-to-point force that relates directly to the intermolecular charge balance of the particle or moiety relative to its environment. Charges arise from two specific types of interactions. First, the valence stability of an atom or moiety in a given environment may favor an unbalanced charge conformation. This is seen with ionizable salts where the electron affinity of a given anion is greater than the electron affinity of the corresponding cation. Hence, the lowest energy conformation results in a charge separation. The energy change between the neutral and the charged form is referred to as the ionization energy.

Coulomb forces also can arise from electron stripping. This occurs when an external force causes the separation of a charge from its neutral location. The charge separation actually results in an increase in the energy state of the system. However, the system delays the return to ground state by the activation energy involved in reversing the charge separation. An example of this would be a material with a low dielectric constant, such as polystyrene, whose electrons are removed from the surface as the result of an implied electromagnetic field resulting in a net static charge. The resistive nature of the material slows electron movement to fill the charge hole, thereby returning to the ground state.

The development of a net charge on the surface of a nanoparticle affects the ion/dipole distribution of the constituents of the solvent (in this case, air or water) immediately adjacent to the nanoparticle. Specifically, a collection of counter ions immediately adjoins the charged surface. The layer of counter ions and the associated net charge, which moves with the Brownian motion of the nanoparticle, is referred to as the Stern layer. If the ions in this layer do not balance the particle's surface charge, the net difference (the Stern potential) then acts upon the rest of the suspension's constituents. The differential movement of the Stern potential within the fluid medium produces an electromagnetic shear force referred to as the zeta potential (ξ). For considerations here, the zeta potential can be generalized to be the net charge of the nanoparticle as presented to the environment. In modeling particle stability or kinetics for larger particles, the displacement of the Stern layer can be ignored. However, for nanoparticles, the presence of the Stern layer may have a significant effect and should be considered integral in the derivation of particle density and volume.

Electrostatic or Coulomb forces generally cause like particles, which tend to acquire like charges, to repel each other. These forces oppose van der Waals force-mediated agglomeration into larger clusters (as described below). While the application of this theory to engineered nanoparticles may be new, engineers have applied the underlying science to water and wastewater treatment processes since at least the 1800s [3]. In the water treatment process of coagulation, operators add chemicals to destabilize colloidal suspensions of naturally occurring nanoparticles. These additives suppress the double-layer charge described above, enabling particles to contact one another and adhere by van der Waals forces. Chapter 7 provides further information on this form of treatment.

6.2.2.2   van der Waals Forces

The van der Waals forces (F_{vdw}) also represent a point-to-point interaction between molecular moieties. They differ from electrorepulsive force in that the charge separation is intramolecular, and therefore the force potential is a fraction of charge per moiety. At the scale of nanoparticles, van der Waals forces are always attractive. They are principally the sum of three component forces: (1) the Keesom force, (2) the Debye force, and (3) the London dispersion force. The Keesom force results from interactions between two permanent dipoles. An example would be the interactions between water molecules or between ionized salts and water molecules. The Debye force represents the interaction between a permanent dipole and an inducible dipole, which results from the electromagnetic field associated with the permanent dipole inducing a charge separation in the transient dipole. In fluid systems, the magnitude of this induction tends to vary in the infrared frequency as the result of molecular vibration of the permanent dipole. An example would be the interactions between water and unsaturated organics, where the water's dipole can induce asymmetric displacement of π-electrons. The London force is the interaction of two induced dipoles that result from the interaction of the electromagnetic fields of two molecules. While this force is universal, it tends to be weaker than the Keesom and Debye forces under typical environmental conditions. Refer to Ackler et al. [4] for examples of application.

The van der Waals forces cause nanoparticles to be attracted to each other as well as to certain other environmental constituents. As a result, nanoparticles can form larger agglomerates. These agglomerates generally tend to be less buoyant and therefore more readily settle out of suspension.

6.2.2.3 Solvency Force

The solvency force (F_s) differs from the electrostatic and van der Waals forces in that it is not a point-to-point interaction. Rather, it is a free energy gradient resulting from the differential energy levels of the pure solvent and the solvent plus the nanoparticle. For example, dispersion of a nanomaterial X in water (hydrosol) with two water binding sites on each nanoparticle requires that the water molecules go from being associated with other water molecules to being associated with the nanoparticles:

$$X + H_2O{\cdot}H_2O \xrightarrow{\ G\ } H_2O{\cdot}X{\cdot}H_2O \qquad (6.8)$$

The net free energy difference (ΔG) between $X + H_2O{\cdot}H_2O$ and $H_2O{\cdot}X{\cdot}H_2O$ is referred to as the free energy of solvation. If the free energy of solvation is thermodynamically advantageous ($\Delta G < 0$), then the material will spontaneously disperse in water. The force component of this energy gradient therefore is the force of solvency. In practice, one can quantify the solvency force by the dispersibility of the material, one of the critical properties of nanomaterials identified in Table 2.2.

6.2.3 IMPLICATIONS OF POLYMORPHISM

The degree of polymorphism also affects the physical and chemical properties of nanomaterials. Polymorphism is the ability of a material to manifest more than one form. As discussed previously, the base molecular structures of almost all nanomaterials are crystalline in nature. Most nanomaterial preparations comprise a distribution of particle sizes as a function of the material's mode of synthesis. This often is referred to as single-component polymorphism.

Another significant form of polymorphism is the interparticle structure of the materials that can form multi-component crystalline phases. For example, carbon nanotubes can form either aligned bundles or tangles referred to as nanoropes. Each form has differing surface properties and electrical densities [5].

A third type of polymorphism occurs when the host nanoparticles condense with guest molecules in heterogeneous structures. Such guest molecules may include solvents, respective counter-valent ions (salts), or other solids (co-crystals). This form of polymorphism often is seen when nanoparticles condense while still in association with their Stern layer constituents as guest molecules. In practice, polymorphism can result in significantly different properties for nanoparticles of the same material. Rudalevige et al. [6] reported this phenomenon for fullerenes, where the crystalline properties of the agglomerated material vary based on the medium from which it condensed.

Because polymorphism can cause variations in physical and chemical properties, care must be taken in extrapolating from the experimental results for a nanomaterial.

6.3 PREDICTING THE BEHAVIOR OF NANOMATERIALS IN THE ENVIRONMENT

The interactions of any given nanomaterial with its environment depend on both the physical and chemical properties described above. All nanomaterials will behave differently because their physical and chemical natures vary with composition and structure. However, by placing the known properties of the materials within an environmental context, it is possible to generally predict a material's transport within the environment and the thermodynamics of potential interactions with the environment.

Because the ultimate purpose for predicting the fate and transport of a material often is to determine the potential for an adverse environmental effect, it is useful to consider the environmental interactions within the context of the risk paradigm. For nanomaterials, this can be divided into three principal considerations:

1. Potential and rate of dispersal or agglomeration in environmental media.
2. Potential and rate of interactions with environmental constituents.
3. Rate and form that a nanomaterial will be presented to an environmental receptor of concern. (Chapters 8 and 9 discuss the potential results of exposure.)

As with any material, nanomaterials will tend toward their equilibrium state ($\Delta G = 0$) within their environment. While this makes it very straightforward to determine the equilibrium conditions for a given situation, complications related to particulate properties can result in significant variability in the transient states. In consequence, it can be difficult to predict the precise kinetics and therefore the time course by which a nanomaterial will transform from the state in which it enters the environment to its ultimate equilibrium state. For example, consider dispersion and agglomeration.

Considerations of dispersion and agglomeration are akin to solubility and vapor pressure for non-nanomaterials, in that they form the basis for predicting the concentrations of materials in environmental media (air or water) relative to the amounts released. However, while vapor pressure and solubility are equilibrium measures, dispersion and agglomeration are dynamic measures. This difference results from the scale of events involved. For example, a small volatile molecule such as vinyl chloride will reach equilibrium vapor pressure very quickly such that the period of disequilibrium becomes insignificant within an environmental context. An aerosol of titanium dioxide in nanoparticulate form, however, may take hours or even days to reach equilibrium. Depending on the nature of the exposure, generalizing equilibrium in such cases may introduce significant uncertainty that may be over- or under-predictive. In risk assessments where assumptions of equilibrium are not appropriate, dynamic prediction methods may need to be applied to develop reasonable estimates of safety. Dynamic prediction differs from equilibrium in that it requires a time-to-event consideration. The changes in the nature of nanomaterial with time are based on the kinetics of important competing reactions that occur as the system moves from a state of disequilibrium, usually at the point of introduction to the environment, to equilibrium. A quantitative approach to dynamic prediction in risk assessment is discussed in the next section.

6.3.1 PREDICTING TEMPORAL REACTION RATES: CHAIN INTERACTIONS

Chemical reaction kinetics is a quantitative generalization between the rate of a reaction going figuratively forward, and the rate of the reaction going backward. Take, for example, the agglomeration of two nanoparticles X:

$$X + X \xrightarrow{\ k_{xx}\ } XX$$

$$\frac{XX \xrightarrow{\ k_{-xx}\ } X + X}{X + X \leftrightarrow XX}$$

$$Rate = -\frac{1}{2}\frac{d[X]}{dt} = k_{xx}[X]^2 - k_{-xx}[XX] \tag{6.9}$$

The accumulation rate of the agglomerate XX is the difference between the rate of agglomeration ($k_{xx}[X]^2$) and the stability of the agglomerate ($k_{-xx}[XX]$). Many of the engineered nanoparticles currently in use, particularly the carbonaceous nanomaterials, form stable aggregates because the combined electrostatic repulsion and energy of solvation cannot overcome the van der Waals forces under typical ambient conditions (i.e., $k_{xx} \gg k_{-xx}$). This allows the following simplification: the rate at which X agglomerates to XX is merely the product of the rate of interaction between Xs and the probability that a given interaction will result in the formation of the product XX.

The rate of interaction between Xs, or the collision kinetics, is governed by the particle size of X and the balance between the system's energy (temperature) and resistance to movement (viscosity). With an estimate of the rate of collision, the rate of product formation can be quantified based on the rate of reaction per collision as follows:

$$k_{XX} = P(r)\frac{2kT}{3\eta} \cdot \frac{2}{r_X} \cdot 2r_X$$

$$= P(r)\frac{8kT}{3\eta} \tag{6.10}$$

where η is the viscosity of the solvent and $P(r)$ is the probability of a reaction resulting in product formation on a per-collision basis.

Because each productive interaction in an agglomeration reaction will increase the particle size by the sum of the two particles, the agglomeration reaction becomes asymmetric very quickly. It must be described as an interaction between unlike particles X and X', where X' is the product of a defined number of agglomeration steps with a rate constant of ($k_{XX'}$):

$$k_{XX'} = P(r)\frac{2kT}{3\eta} \cdot \left(2 + \frac{r_X}{r_{X'}} + \frac{r_{X'}}{r_X}\right) \tag{6.11}$$

Because of the relatively large size of nanoparticles (compared to typical molecules), the asymmetry between the initial particle radius r_X and the radius of the agglomerated particle $r_{X'}$ grows very large as the result of a relatively small number of agglomeration reactions. Hence, even if there is no change in the probability of agglomeration $P(r)$, the reaction rate will change significantly with time and independent of relative concentrations. This is further compounded by the large number of coupled agglomeration reactions involved ($X + X$, $X + XX$, $XX + XX$, $X + XXX$, $XX + XXX$, ...) in the evolution of suspended nanoparticles into large particles that cannot remain in suspension.

Fortunately, the need to estimate overall reaction rates with time-variable reaction constants is not unique to nanomaterials. It was a problem first encountered in nuclear physics in solving multi-stage chain reactions. Nuclear physicists overcame this problem using multiple stochastic reaction simulations with randomized iterations, also referred to as Monte Carlo simulation. Gillespie [7] proposed one approach, originally developed to predict water droplet aggregation in clouds, that is particularly applicable to the agglomeration of nanomaterials in suspension. It is a sequential stochastic simulation that predicts the concentration of various defined products/ reactants by determining the probability of the most likely reaction $(P(\mu))$ to occur between time t and time t+τ based on the competitive values for the respective reaction rates (k') specific for time t $(P(\tau,\mu))$.

The stochastic probability model divides the reaction probability into two probabilities: (1) the independent probability of any reaction occurring in the duration of τ $(P_1(\tau))$, and (2) the dependent probability of a specific reaction (μ) occurring given a specific value for τ $(P_2(\mu|\tau))$:

$$P(\tau,\mu) = P_1(\tau) \cdot P_2(\mu|\tau) \tag{6.12}$$

The infinitesimal of the probability, $P(\tau,\mu)$ $d\tau$, represents the probability at time t that the next reaction will occur in the differential time interval of $t+\tau$ to $t+\tau$ $d\tau$. For any specific reaction, μ, the probability of co-occurrence within $d\tau$ if the product of the rate of diffusive interaction $(k_{D\mu})$ and the number of distinct reactant combinations found present at time $t(h_\mu)$ is as follows:

$$P(\mu)d\tau = h_\mu \cdot k_{D\mu}\, d\tau \tag{6.13}$$

The value of h_μ can be determined by the nature of the reaction as to how the respective reactant concentrations change with production of the product (Y) with each reaction event μ using the following relations:

$$X + X \rightarrow Y \quad h_\mu = \frac{[X] \cdot ([X]-1)}{2} \tag{6.14}$$

$$X + X' \rightarrow Y \quad h_\mu = [X] \cdot [X']$$

Hence, the probability of a given reaction occurring in the time period of t+τ to t+τ dτ is a function of the independent probabilities of no reaction occurring ($P_0(\tau)$), and the probability that reaction μ will occur ($P(\tau,\mu)$) as follows:

$$P(\tau,\mu)d\tau = P_0(\tau) \cdot P(\mu)d\tau$$

$$= P_0(\tau) \cdot \left(h_\mu \cdot k'_\mu \, d\tau \right) \tag{6.15}$$

$P_0(\tau)$ is the integration of the negative likelihood of a reaction occurring within the time period τ. Because μ is the most likely reaction at time t and defines the duration of the time-step τ, it is the most likely and only reaction to occur within the defined time-step. To identify and define reaction μ, the standard limit formula can be applied for all possible reactions ($\mu = \{1,\ldots, M\}$) at time t to provide a relation for $P_0(\tau)$ as follows:

$$P_0(\tau) = \left(1 - P(\tau)\right) = e^{-\sum_{\mu=1}^{M} h'_\mu k_\mu \tau} \tag{6.16}$$

Substituting this into the previous probability relationship provides an expression for the probability of a reaction occurring within the prescribed τ as follows:

$$P(\tau,\mu) = h_\mu k'_\mu \cdot e^{-\sum_{i=1}^{M} h_\mu k'_\mu \tau} \tag{6.17}$$

Going back to the original defining probability where $P_1(\tau)$ is defined as the probability of any specific reaction from 1 to M occurring in the duration of τ at time t, it can now be defined as the summation of $P(\tau,\mu)$:

$$P_1(\tau) = \sum_{i=1}^{M} P_1(\tau,\mu_i)$$

$$= \sum_{i=1}^{M} (h_i \, k'_i) \cdot e^{-\sum_{i=1}^{M} h_i \, k'_{Di} \tau} \tag{6.18}$$

The derivation of τ at time t is not absolute, but rather a value from a distribution of time intervals based on the respective reaction rates for the M reactions possible, and hence can be simulated as follows:

$$\tau = \frac{1}{\sum_{\mu=1}^{M} (h_\mu k'_{D\mu})} \cdot \ln\left(\frac{1}{r_1}\right) \tag{6.19}$$

where r_1 represents a random variable from a uniform distribution of $\{0,\dots, 1\}$.

To complete the expression for the defining probability, a relation for $P_2(\mu|\tau)$ can be developed by substituting the above equation into the defining probability as follows:

$$P_2\left(\mu \mid \tau\right) = \frac{P\left(\tau,\mu\right)}{\displaystyle\sum_{\mu=1}^{M} P_1\left(\tau,\mu\right)}$$

$$= \frac{h\mu \cdot k_\mu}{\displaystyle\sum_{\mu-1}^{M}\left(h_\mu k'_\mu\right)\cdot e^{-\sum_{\mu-1}^{M} h_\mu k'_\mu \tau}} \tag{6.20}$$

Again, $P_2(\mu|\tau)$ is not an absolute, but rather a probability distribution function. In this case, the probability of a reaction occurring is based on the relative reaction rate. The solution for the distribution therefore can be simulated using a second uniform random variable (r_2) and solving for μ in the relation whereby:

$$\sum_{i=1}^{\mu-1}\left(h_i k'_{Di}\right) < r_2 \sum_{i=1}^{M}\left(h_i k'_{Di}\right) \le \sum_{i=1}^{\mu}\left(h_i k'_{Di}\right) \quad r_2 \in Uniform\{0 < r_2 < 1\} \tag{6.21}$$

The order of the summation is irrelevant. Therefore, this relation can be solved mathematically by successive summations until the following condition is met:

$$r_2 \sum_{i=1}^{M}\left(h_i k'_{Di}\right) - \sum_{i=1}^{n}\left(h_i k'_{Di}\right) < 0 \quad n \in \{1,2,3,\dots,M\} \tag{6.22}$$

When this is satisfied, the value of μ is therefore that corresponding to the prior value of i (μ_{i-1}).

Although theoretically complex, this approach allows for the prediction of the rate of aggregate formation regardless of the number of separate types of reactions or the number of intermediates involved. It also foregoes the need to solve a generalized master equation by considering all potential interactions simultaneously. It is very powerful; however, it is also very computationally intensive.

An example for the application of Gillespie's model to predict the collision kinetics for an agglomeration reaction is illustrated in Figure 6.5. As expected, the lower the probability of product formation, the longer the process of chain reaction agglomeration. It is interesting that the uncertainty also increases. This uncertainty is not the result of prediction (experimental) error, but rather represents differential reaction pathways and is a true measure of the variance expected if such a reaction were repeated an infinite number of times. This again is the result of the large number of potential intermediates possible in the aggregation between the slowest linear aggregation pathway ($X + X$, $XX + X$, $XXX + X$, ...) and the fastest geometric

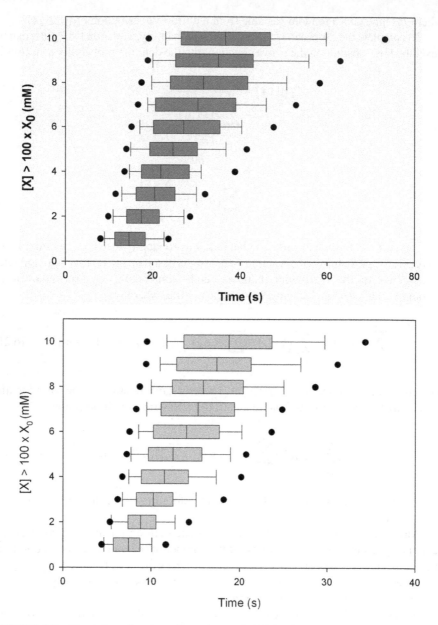

FIGURE 6.5 Examples of projected reaction probabilities based on stochastic kinetics: (a) representation of variability in product formation for the agglomeration of a 10-nm particle with $P(r) = 0.1$; (b) example of projected probability of agglomeration at differing particle size at an assumed $P(r)$ of 0.5

aggregation pathway ($X + X$, $XX + XX$, $XXX + XXX$, ...). Using this approach, not only is the range recognized, but also the relative probability, which is a function of the relative collision kinetics of the intermediates, is retained.

6.3.2 PREDICTING TEMPORAL REACTION RATES: ESTIMATING PARTICLE AFFINITIES

In addition to a method to determine the time course of the collision/diffusion kinetics, prediction of the fate of nanomaterials requires derivation of the probability of a reaction resulting in the formation of a product per collision event, $P(r)$. Experimentally, this is reasonably easy to determine within the confidence of the collision kinetics as the ratio of observed product formation given the determined rate of collision:

$$X + X \rightarrow Y \quad P(r) = \frac{k_{x'}}{k_{Dx}} \tag{6.23}$$

where k'_{Dx} is the rate of collision based on diffusion and $k'_{x'}$ is the rate of product formation. Deriving $P(r)$ from thermodynamic principles is difficult because of the number of competing forces and from the limited knowledge regarding near-body interactions in solution. Hence, the methods described below should only be considered a means of estimation.

It is generally true that the more thermodynamically advantageous a reaction, the more likely it is to occur, and therefore the faster the rate of product formation. With respect to the agglomeration of nanoparticles, product formation occurs when the forces of attraction outweigh the forces of repulsion. This summation, however, is not straightforward because the molecular force fields around each nanoparticle vary with distance from the particle. The energy required to overcome these force fields depends on the kinetic energy of the particles, which is neither constant nor uniform.

Derivation of predictive values for the free energy of solvation — and its inverse, the free energy of precipitation — takes into account the affinity of the solvent (in this case, air or water) for the solute relative to the affinity of the solute particle for other solute particles. These affinities are chemical specific. However, it is possible to generalize the interactions of a nanoparticle with its solvent medium.

Consider an example of a nanoparticle introduced to an aqueous medium:

- If the nanoparticle's surface affinity for like nanoparticles is low relative to the affinity for the water molecules, then the material will disperse.
- If the nanoparticle has a low affinity for like nanoparticles but its affinity for polar water molecules is insufficient to overcome the water–water affinity, then the material will be hydrophobic and will not disperse in water but will disperse in nonpolar environments at the solvent interface.
- If the nanomaterial has a high affinity for like nanoparticles, the material will not disperse in either aqueous or nonaqueous environments.

These situations are never absolute. In general, the stronger the affinity of the nanoparticle for water, the higher the equilibrium concentration — and vice versa. (Recall that if the free energy of solvation is less than zero, then a material will disperse spontaneously in water.)

Dispersion in air (aerosol) differs from hydrosol formation principally because (1) the fluid medium has a lower density and higher particle velocities; (2) the medium has a low dipole moment; and (3) the medium has a low dielectric constant. Therefore, the primary factors in air dispersion are particle size and inter-particle affinities that are related to inducible net zeta potential in air.

In both cases — dispersion in water and dispersion in air — the fate of the nanoparticle results from the interplay of competing interactions at the nanoparticle interface. To predict the probability of agglomeration and thereby the stability of the nanomaterial, the force fields at this interface must be described in thermodynamic terms that then can be converted to a probability density function.

6.3.3 Nanoparticle Affinity and Inter-Particle Force Fields

Interactions between nanoparticles and environmental constituents such as fluid media are expected to result predominantly from Coulomb (electrostatic) forces and van der Waals interactions. That is not to say that nanomaterials will not undergo covalent reactions within the environment. An example of such a reaction is the application of zero-valent iron in groundwater remediation where the iron nanoparticles undergo direct redox reactions with groundwater contaminants [8]. However, this is the exception and specific to the type of nanoparticles involved. Coulomb forces will occur in any situation where the particle/medium system permits the formation of a charge imbalance. van der Waals interactions are universal to nanoparticles and will differ among type only with regard to their magnitude.

6.3.3.1 Coulomb Energy

In agglomeration reactions, the Coulomb force is almost always repulsive. This occurs because it is most common that like particles in the same medium will acquire the same type of charge, although the charge density may vary with the particle size. Charges can arise as the result of charge separation producing a dipole situation, but unlike molecular dipoles, this is usually aligned between the outside surface of the particle and its interior. As such, steric hindrance inhibits differential charge interactions. The potential energy $(E(C)_{xx'})$ arising from the Coulomb forces between the two particles, X and X', can be defined as follows:

$$E(C)_{xx'} = \frac{q_x \cdot q_{x'}}{4\pi \cdot z \cdot \varepsilon_0 \cdot \varepsilon_s} \tag{6.24}$$

where q is the net particle charge on X or X', ε_0 is the electric constant (8.85×10^{-12} $C^2 \cdot N^{-1} \cdot m^{-2}$), ε_s is the dielectric constant of the medium, and z is the particle separation [9]. This can be optimized for the interaction between two spheres as follows [10]:

$$E(C)_{xx'} = \frac{q_x \cdot q_{x'} e^{-\kappa z}}{4\pi \cdot \varepsilon_0 \cdot \varepsilon_s \cdot z_0 \cdot (1 + 2\kappa z)} \tag{6.25}$$

where κ is the inverse Debye screening length (≈ 1.43 nm).

A positive energy is repulsive; a negative energy is attractive.

6.3.3.2 van der Waals Energy

The net van der Waals force is a balance of weak attractive and repulsive interactions between either nanoparticle surfaces or the nanoparticle surface and other medium constituents. Over molecular distances the net force is always attractive and exothermic, with the change in free energy being the result of the enthalpy of adsorption (^-H_a). (Over atomic distances, the force is always repulsive.) This balance can be approximated by the Lennard-Jones (12-6) relation [11], where intermolecular potential energy ($E(w)$) is given by:

$$E(w)_{xx'} = 4 \; H_a \left[\left(\frac{z_0}{z} \right)^{12} - 2 \cdot \left(\frac{z_0}{z} \right)^6 \right] \tag{6.26}$$

where z is the distance between two particles, and z_0 is the most thermodynamically favorable distance at which $E(w)$ is equal to ^-H_a. The derivation of the Lennard-Jones relation comes from the differences between the attractive forces that vary with the 6th power of the inverse distance, and the repulsive force that varies with the 12th power. Note that the parameters represent the summation of paired potentials across the interacting surface. Therefore, the values for $-H_a$ and z_0 will not be the same in an agglomerate such as a nanoparticle, as they would for the individual molecular or atomic constituents.

The relationship changes when dealing with a molecular/nanoparticle interaction. This is because the potential is based on the summation of paired interactions of one body acting on multiple single points. As a result, the relation changes from a (12-6) to a (9-3) [12] as follows:

$$E(w)_{xx'} = 4\pi n z_0^3 \; H_a \left[\frac{1}{45} \left(\frac{z_0}{z} \right)^9 - \frac{1}{6} \left(\frac{z_0}{z} \right)^3 \right] \tag{6.27}$$

where n is the number of binding sites upon the nanoparticle. Examples of the differential relations are provided in Figure 6.6 for C60 fullerene-fullerene [A] and C60 fullerene and water [B].

Determinations of the van der Waals energy are difficult, particularly for opaque materials. However, the energy can be predicted for a binary system of two like particles (x) in a solvent (s) based on the Hamaker constant (A). The Hamaker constant can be estimated within a given system based on the reference dielectric constant in a vacuum ($\varepsilon_{0,n}$) using the Tabor Winterton approximation [13] as follows:

$$E(w)_{xsx} = -\frac{A_{xsx}}{12\pi \cdot z^m}$$

where :

$$A_{xsx} = \frac{9 x 10^{15} \pi \hbar}{8\sqrt{2}} \cdot \frac{\left(\varepsilon_{0,x}^4 - \varepsilon_{0,s}^4 \right)^2}{\left(\varepsilon_{0,x}^4 + \varepsilon_{0,s}^4 \right)^{3/2}} \tag{6.28}$$

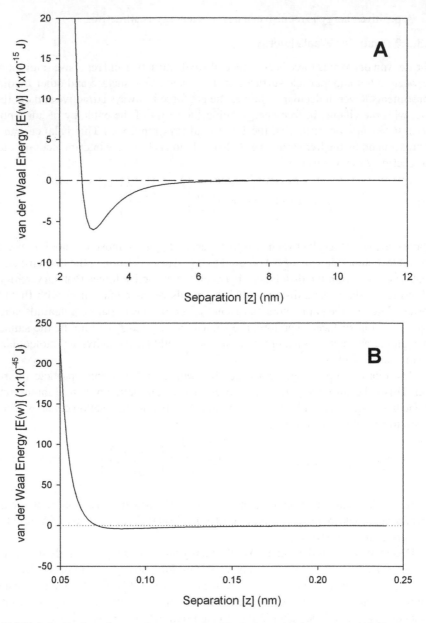

FIGURE 6.6 Projected examples of van der Waals force (A) between two fullerene molecules and (B) between a fullerene and water molecule. Projections parameterized based on the observations of Chen and Elimelech [14, 15] and Labille et al. [16].

where ℏ is Planck's constant, and m is a geometric constant that can be applied using the semi-empirical values in Table 6.2.

TABLE 6.2
Empirical Coefficients for M in the
Tabor Winterton Approximation

Geometry	M
Molecular point-to-point	6
Two-plane parallel bodies	2
Two spherical particles	1

Note: Table data taken from French [1].

6.3.4 Prediction of Probability of Product Formation

Predicting the kinetics of nanoparticles' mobility in the environment requires the quantification of the probability of product formation relative to the collision kinetics. The enthalpy of adsorption is the critical factor in predicting the probability of product formation. For simplicity, the derivation begins with an assumption of uniform dispersion within an aqueous medium. While this assumption is not necessary to validate the solution, it removes considerations of steric hindrances while presenting a more intuitive model.

From this initial dispersed situation, the free energy of a single agglomeration reaction has three components, which sum as follows:

$$
\begin{aligned}
&(1)\ 2X \cdot H_2O \leftrightarrow 2X + 2H_2O \\
&(2)\ H_2O + H_2O \leftrightarrow H_2O \cdot H_2O \\
&\underline{(3)\ X + X \leftrightarrow X \cdot X} \\
&(4)\ 2X \cdot H_2O \leftrightarrow X \cdot X + H_2O \cdot H_2O
\end{aligned}
\tag{6.29}
$$

For simplicity, it can be assumed that the free energy of $X \cdot H_2O$ is independent of the number of water molecules present, and the free energy of $X \cdot X$ is independent of the number of nanoparticles previously combined.

Every interaction will result in either the formation of a product $(X \cdot X)$ or the elastic rebound of the reactants $(X \cdot H_2O)$. Hence, the expression for the probability of outcome per collision can be described as follows:

$$
P(X \cdot X) + P(2X \cdot H_2O) = 1
\tag{6.30}
$$

Because van der Waals energy is always negative, the probability of $(X \cdot X)$ will be 0.5, provided that no steric hindrance or electrostatic potential inhibits the agglomeration. Considering this is a simple particle agglomeration, it can be assumed that steric hindrance is not an issue. Therefore, as the electrostatic repulsion increases, the probability favors the $P(X \cdot H_2O)$ over the $P(X \cdot X)$, and vice versa. Using DLVO kinetics, the aggregation efficiency (α), sometimes referred to as the inverse stability $(1/W)$, can be expressed as follows:

$$\alpha_{xx} = \frac{e^{\frac{E(w)_{xx}}{KT}}}{e^{\frac{E(w)_{xx}+E(C)_{xx}}{KT}}}$$

where

$$E(w)_{xsx} = -\frac{A_{xsx}}{12\pi \cdot z^m}$$

and   (6.31)

$$E(C)_{xx} = \frac{d_x^2}{4\pi \cdot z \cdot \varepsilon_0 \cdot \varepsilon_s}$$

and

$$\varepsilon_s = \sum \varepsilon_i \cdot [i] \cdot MR_i$$

where MR is the molar refractivity (l/m) of constituent i.

Because the probabilities are based on the energy balance at the point of collision, they are concentration independent. The point of collision, defined as the effective particle radius (z) in the interaction model, represents the distance where the kinetic energy of the particles is equal to the repulsive forces. Therefore, the probability of interaction is equal to the agglomeration efficiency:

$$\alpha_{xx} = P(X \cdot X)$$

and   (6.32)

$$P(X \cdot H_2O) = 1 - \alpha_{xx}$$

6.3.5 SUMMARY

The approach described above can be extremely useful in assessing the possible risks from nanomaterials because it permits the prediction of significant nonequilibrium behavior based on measurable physical properties. Simple qualitative assessment will enable a determination of the stability of the nanoparticles. Detailed quantitative assessment will allow the prediction of the particles' behavior, and thereby the extent of potential distribution within the environment.

Consider the example of dispersed C60 fullerene. Materials such as carbon nanotubes and fullerenes are not stable in the environment and will agglomerate under conditions where the van der Waals attraction can overcome electrostatic forces. When dispersed as either a hydrosol or aerosol, usually as the result of mechanical agitation, carbonaceous nanoparticles immediately begin to agglomerate, forming larger and larger super-particles. The rate of agglomeration is a function of the immediate concentration of the materials. Because the particles are subject to diffusion, a continuous release results in equilibrium of spatial size distribution that is the result of the disequilibrium of the materials themselves.

In this example, a suspension of dispersed C60 fullerene in a concentration of 0.1 milligram per liter (mg/L) is discharged at a rate of 1 liter per minute (L/min) to a flowing creek system (0.1 m/sec). The Hamaker constant for fullerene (A_{FWF}) is approximately 6.7×10^{-21} J [14]. Assuming a water hardness of approximately 80 mg/L $CaCO_3$, yields a $P(F \cdot F)$ of approximately 0.63.

If a hypothetical regulatory limit of 1 µg/L of particles with an apparent diameter less than 250 nm is imposed, it is possible to predict the extent to which the creek is potentially out of compliance. Assuming instant chemical equilibrium, the material would be expected to be entirely agglomerated and therefore no amount of release would result in noncompliance. Under a conservative assumption of no agglomeration, the reach of the river that would be considered out of compliance would extend from the point of release until the point where diffusion and mixing diluted the fullerenes sufficiently. Compliance would require dilution by a factor of 100:1. Alternatively, one can predict the extent of agglomeration as a basis for assessing compliance. The competitive stochastic modeling allows for an incremental analysis of the size distribution of the fullerenes within the river (Figure 6.7). The agglomeration model indicates that compliance would be achieved within 9 m downstream of the discharge.

6.4   RESEARCH RESULTS

Experimental studies regarding the fate and transport of nanomaterials in the environment currently fall into two broad categories: behavior in aqueous systems and movement through porous media. As discussed further in Chapter 7, the behavior of nanomaterials in water and wastewater has been investigated as a basis for evaluating the effectiveness of various treatment technologies such as coagulation/flocculation and filtration. Migration of nZVI through the subsurface has been studied with regard to its application as a groundwater treatment technology, as described in Chapter 10. The findings of this and other research relevant to characterizing the fate and transport of nanomaterials in surface water, sediments, and groundwater are discussed below.

Much research is underway to characterize the behavior of nanoparticles under "environmentally relevant conditions." This research shows that a range of variables complicate the behavior of nanomaterials in the environment. These variables include the pH and ionic strength of the aqueous solution, the presence of dissolved organic matter, and the organic carbon content and grain size of the soil. Modifications in the physicochemical properties of the nanoparticles, either engineered or occurring upon release to the environment, may lead to unpredictable transport behavior in surface water and groundwater. After assessing the mobility of eight different nanomaterials in a porous medium in laboratory experiments, researchers at Rice University concluded that "The differences in the environmental transport properties for these nanomaterials underscores the need to address environmental impacts of nanomaterials on a case-by-case basis" [17]. The characteristics of both the nanomaterial and the environmental system will affect the fate and transport of nanomaterials

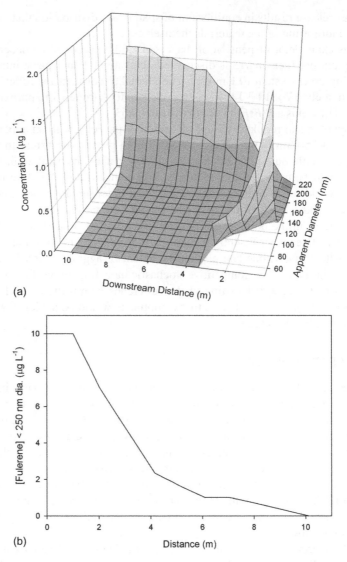

FIGURE 6.7   Simulation of a fullerene release into a flowing creek: (a) distribution of apparent particle diameter with downstream distance. Projection based on competitive stoichiometric analysis as a resolution of ±10 pmol. (b) Projected concentration of fullerene in the creek below a theoretical compliance limit of 1 μg/L of fullerene under 250-nm in apparent diameter.

6.4.1   SURFACE WATER AND SEDIMENT

Research on nanotubes in aqueous systems has been well documented. Researchers at the Georgia Institute of Technology have investigated the aqueous stability of multi-walled carbon nanotubes in the presence of natural but undefined organic matter. Because these nanomaterials are hydrophobic, they would be expected to agglomerate and settle from the water column. However, in the presence of natural organic matter at varying concentrations in laboratory solutions and at background concentrations in actual river water samples, the carbon nanotubes remained in a stable, dispersed state for more than one month. The natural organic matter appeared to be a better stabilizing agent than sodium dodecyl sulfate, a surfactant often applied in industrial processes to stabilize carbon nanotubes [18].

Studies by Chen and Elimelech [14] found a similar response with fullerenes where a concentration-dependent inverse effect of α was observed. The presence of excess humic acid appeared to increase the electrostatic hindrance to agglomeration. This increased the critical coagulation concentration (i.e., the minimum concentration of a coagulant necessary to suppress the double-layer charge and allow particles to agglomerate) from 8.0 to 19.4 mM $MgCl_2$. This result is consistent with and predictable based on DLVO kinetics. However, Chen and Elimelech also observed that high calcium concentrations above the critical coagulation concentration (40 mM) increased the rate of agglomeration of C60 fullerenes relative to that in an untreated suspension. Although not addressed in the article, this higher rate of agglomeration can be accounted for by changes in collision kinetics relative to the humic acid plus fullerene $P(r)/\alpha$ values.

Nowack reviewed the literature on C60 fullerenes naturally occurring in ancient geologic materials and concluded that "the stability of fullerenes under geologic conditions for hundreds of millions of years shows that they are truly recalcitrant in the environment." While pure fullerenes are nearly insoluble in water, under certain conditions fullerenes will form polymorphic hexagonal unit cell agglomerates in water referred to as nano-C60. These agglomerates, approximately 25 to 500 nm in size, carry a strong negative charge [19]. The physical and chemical properties of the agglomerate nano-C60, such as color, hydrophobicity, and reactivity, are significantly different as the result of the differing crystalline structure that can be manipulated by controlling the solution pH and the rate at which water is added. The critical coagulant concentration for these nano-C60 agglomerates is in excess of 500 mM NaCl, indicating a significant increase in electrostatic hindrance to agglomeration relative to the individual fullerenes [20]. Such dramatic deviations in surface properties as the result of polymorphism further emphasize the importance of nanomaterial characterization in predicting environmental fate and transport.

Researchers at the Georgia Institute of Technology also have studied the photochemical reactions that affect fullerenes in an aqueous system. In their studies, they prepared various polymorphic agglomerate forms of fullerenes in water: nano-C60 suspension prepared by solvent exchange, nano-C60 suspension prepared by sonication, C60 stabilized in water with polymers and surfactants, and C60 stabilized in water by natural organic matter. They evaluated the photochemical reactivity of the various dispersed forms of C60 by measuring the production of reactive oxygen species (specifically the singlet oxygen and superoxide radical anion). The researchers found that the photochemical reactivity of the fullerenes, or the ability of the particles to mediate energy and electron transfer, was a function of the polymorphic nature of the nanomaterial and the characteristics of the stabilizing molecules [21].

Other researchers at Purdue University, funded by the U.S. EPA's National Center for Environmental Research (NCER) for the period from 2007 to 2009, are studying the photodegradation of fullerenes and single-walled carbon nanotubes. The research team will irradiate solutions or suspensions of fullerenes and nanotubes and monitor the subsequent loss rate and product formation rate, as well as the spectroscopic and microscopic characteristics of the solutions. The water will contain various naturally occurring substances such as carbonates, humic acids, and oxygen [22]. No progress report was available at the writing of this book

Few publications on the fate and transport of nanomaterials in sediments were available at the time this book was written. Researchers at the University of South Carolina at Columbia currently are investigating the "chemical and biological behavior of carbon nanotubes in estuarine sedimentary systems" under a grant funded by NCER for the period from 2004 to 2007. The objective of the research is to evaluate the potential for single-walled carbon nanotubes to "be transported, accumulate and cause deleterious effects within estuarine environments" [23]. The initial research results showed the effect of salinity on agglomeration of nanotubes in the water column. Single-walled carbon nanotubes with an average particle size of 200 to 250 nm could exist in stable colloidal suspensions at neutral pH in solutions of low ionic strength (10 mM). Within minutes after increasing the salinity to over 5 parts per thousand, the nanotubes formed large flocs with an average particle size greater than 2 μm. Subsequent research, as yet unreported in the literature, has focused on the distribution of single-walled nanotubes in sediments and in sediment-ingesting benthic invertebrates [24].

A second research group at the University of Michigan has received funding from NCER to evaluate the dispersion states of carbon nanotubes under typical environmental conditions and to evaluate the transport behavior of these nanomaterials in different types of soil and sediment media [25]. However, no progress reports were available for this research project as of this writing.

6.4.2 GROUNDWATER

Various studies have been performed to evaluate the mobility of nanomaterials in porous media. These include studies of the fate and transport of nanomaterials in certain wastewater treatment processes, such as filtration, which provide insights into the potential movement of nanoparticles through groundwater.

As discussed further in Chapter 7, researchers have conducted laboratory experiments to empirically derive the attachment efficiency, α, which is the ratio of the rate of particle deposition to the rate of particle collisions with a filter medium. The attachment efficiency is a single parameter that provides a measure of the various forces acting on a particle as it passes through a porous medium, such as the van der Waals forces, electrostatic repulsive or Coulomb force, and force of solvency, described previously in this chapter.

Lecoanet, Bottero, and Wiesner conducted column experiments to quantify the mobility of eight different manufactured nanomaterials in a porous medium of glass beads, which they indicated would be representative of a sandy groundwater aquifer [17]. The nanomaterials tested included two sizes of silica, titanium dioxide, ferroxane, alumoxane, fullerol (hydroxylated C60), nano-C60, and surface-modified single-walled carbon nanotubes. The results indicated that different forms of nanoparticles with the same composition have different mobilities. For example, of the carbon-based particles tested, single-walled carbon nanotubes and fullerols passed through the porous medium more rapidly than the colloidal nano-C60 agglomerates. The solubilized forms of the particles are more mobile than the suspended form. The researchers identified the need for additional studies to evaluate the factors in natural systems that may increase the mobility of nanoparticles in the environment. These potential factors include naturally occurring polyelectrolytes such as fulvic and humic acids that might affect the surface properties of the nanoparticles, and light- or bio-activated functionalization that would alter the solubility of the nanoparticles [17].

Lecoanet and Wiesner [26] conducted additional experiments to evaluate the effect of flow velocity on the deposition of various nanoparticles in a porous medium. They evaluated two types of fullerenes (fullerol and nano-C60 agglomerates), surface-modified single-walled nanotubes, and two mineral oxides (silica and titanium dioxide). Silica particles showed very little removal and very small variability in changes in flow rate. The passage of titanium dioxide through the porous medium was proportional to the flow rates. However, the affinity of the fullerenes and carbon nanotubes was insensitive to flow rate, which the researchers attributed to the larger Hamaker constants for these nanoparticles. They also noted that at the higher flow rates, the fullerenes exhibited increased deposition after passage of the first pore volume. This suggests that previous fullerene deposition can facilitate subsequent deposition, most likely the result of fullerene-fullerene interactions [26].

The mobility of nZVI and other metallic nanoparticles in aquifer systems has been widely studied because of the potential application of these materials in groundwater remediation. Based on its high redox potential, nZVI can facilitate reductive dechlorination of organic contaminants in groundwater and immobilize heavy metals through cation reduction. However, its effectiveness in groundwater remediation depends on its ability to remain dispersed in the groundwater and not to agglomerate with itself or bind to soil particles. A suspension of nZVI must resist initial agglomeration so it can be delivered effectively through the porous medium to the groundwater contaminants. As discussed in Chapter 10, the fate and transport of nZVI injected into an aquifer depend on the characteristics of the reagent and on the aquifer characteristics, including the flow of groundwater, geochemistry, and the nature of the aquifer materials. Oxidation of nZVI, agglomeration, and attachment to soil grains can occur rapidly. Depending on the setting, nZVI can migrate only a few meters to tens of meters from the injection site, and its reactivity lasts on the order of weeks to months.

Phenrat et al. [27] performed laboratory experiments to evaluate the effect of dispersion concentrations and magnetic attractive forces on the agglomeration rates for various aqueous nZVI dispersions. They concluded that because of the rapid agglomeration of nZVI into micrometer-sized particles, adsorbed polymers are required to modify the surfaces of the nZVI particles and thereby prevent agglomeration and enhance mobility. A decrease in magnetization would also improve the stability of nZVI dispersion [27].

Dunphy Guzman et al. [10] conducted laboratory experiments to evaluate the effect of pH on the agglomeration and transport of titanium dioxide in a porous medium. They reported that pH affects the surface charge of the titanium dioxide, as well as the size of the agglomerated particles. More than 80% of the suspended particles and agglomerates were mobile over the pH range of 1 to 12. The mobility of the particles was not affected except around the pH of zero point charge (pH = 5.9); at this pH, the agglomerates settled out of suspension. However, the researchers noted that this initial mobility might allow nanoparticles to cross redox zones and to move to regions of different solution chemistry or surface charge, resulting in uneven distributions.

Elliott and Zhang [28] performed a field demonstration to evaluate the effectiveness of nanoscale bimetallic particles (Fe/Pd) in degrading chlorinated hydrocarbons. They injected the nanoparticle suspension into a test area of a well-characterized chlorinated hydrocarbon plume and monitored TCE, total iron, and dissolved iron concentrations. They experienced minimal clogging of the injection well and noted that the nanoparticle plume traveled at an apparent velocity of 0.8 m/d, exceeding the natural seepage velocity of 0.3 m/d. They concluded that recirculation of the groundwater prior to injection and dispersion of the nanoparticles had contributed to this latter effect [28]. The movement of nanoscale particle plumes at velocities faster than the natural migration rate of the contaminant plumes has been observed at other sites [29].

While past research has focused on the behavior of nanoparticles in aquifers and other saturated porous media, ongoing research projects are looking at the behavior of nanoparticles in both saturated and unsaturated soils. The NCER has funded research projects on the following topics:

- Agglomeration, retention, and transport behavior of manufactured nanoparticles in variably saturated porous media [30]
- Fate and transport of carbon nanomaterials in unsaturated and saturated soils [31]

The 2006 progress report is available for the latter project, which focuses on nano-C60 and fullerol. During the first year of the project, researchers at the Georgia Institute of Technology conducted 40 column experiments to evaluate the effects of flow rate, soil particle size, and influent concentration on transport of nano-C60 in water-saturated sand or glass beads. They found that the rate of attachment of nano-C60 to the porous medium depended on the number of available deposition sites. At low ionic strength, nano-C60 behaved like a non-reactive tracer and passed through the sand, consistent with the behavior predicted by DLVO theory. The second year of research will look at transport in a natural medium, in the presence of surfactants and dissolved organic matter [31].

6.5 CONCLUSIONS

Laboratory and field studies to date show how nanomaterials interact with each other and with other environmental constituents to affect their fate and transport. The physicochemical properties of the nanoparticles, either as engineered or as modified by reactions in the environment, may lead to transport behavior that is difficult to model precisely. Environmental variables such as the pH and ionic strength of the aqueous solution, the presence of dissolved organic matter, and the organic carbon content and grain size of the soil strongly affect the migration and disposition of nanomaterials. Thus, as with many environmental questions, the fate and transport of nanomaterials must be considered on a case-by-case basis.

The fate and transport of nanomaterials in the environment depend on physical and chemical considerations different than those applied for simple molecular chemicals. The environmental stability of nanoparticles is not as dependent on redox-type reactions, but rather on physisorption properties that permit them to interact with environmental constituents at low energies. Furthermore, because particle sizes and related properties continue to change after a release due to agglomeration and other environmental reactions, assumptions of equilibrium may be highly misleading, particularly when multiple reaction products or intermediates are possible. It therefore is important to consider the kinetics of the nanoparticle interactions within the context of the time course of concomitant environmental interactions involved in fate and transport.

The models provided in this chapter represent initial steps in developing the means that can be applied to predict the behavior of specific materials within the environment, and thereby to determine the potential for an adverse effect. They are the logical replacement for properties such as solubility, Henry's Law constants, and partition coefficients that are the basis for *a priori* evaluation of the environmental fate of smaller and more typical environmental contaminants. The derivations in this chapter also indicate the need to broaden our understanding of the physical properties of nanomaterials. Properties such as zeta potentials and dielectric and Hamaker constants will become critical to adequately predict fate and transport, thereby allowing for due consideration and precaution in the introduction and handling of nanomaterials in the environment.

6.6 LIST OF SYMBOLS

A	Hamaker constant
$E(C)_{XX'}$	Potential energy arising from the Coulomb forces between two particles
$E(w)_{XX'}$	Intermolecular potential energy
$F\!\downarrow$	Vector settling force
F_T	Total force field
F_R	Electrostatic repulsive (Coulomb) force
F_{vdw}	van der Waals forces
F_s	Forces of solvency
g	Gravitational constant
ΔG	Net free energy difference
h	Linear measure of particle separation
ΔH_a	Enthalpy of adsorption
i	Constituent index
k	Boltzmann constant
k_{XX}	Collision rate constant for two identical entities
k_{as}	Apparent solubility
k'	Reaction rate
MR	Molar refractivity
n	Number of binding sites on the nanoparticle

P Probability
q Net particle charge
r Effective particle radius
s Solvent
t Time
T Absolute temperature
W Stability coefficient
X Hypothetical nanoparticle reactant
$[X]_f$ Concentration of nanoparticle in sol form
$[X]_s$ Concentration of nanoparticle in solid (non-sol) form
$[X]_{aq}$ Aqueous concentration
v_x Velocity of X
V_p Particle Volume
V_x Volume
Y Hypothetical product
z Particle separation
z_0 Distance of minimal repulsion
α Aggregation efficiency
K Inverse Debye length
η Viscosity
ρ_{aq} Aqueous density
ρ_f Fluid density
ρ_x Particulate density
ζ Zeta potential
τ Time-step for the duration of a reaction
μ Potential reaction index
ε_0 Vacuum reference dielectric constant
ε_s Dielectric constant of the medium
h_μ Number of distinct reaction combinations for reaction μ
$\hbar$ Planck's constant

REFERENCES

1. Nurmi, J.T., P.G. Tratnyek, V. Sarathy, et al. 2005. Characterization and properties of metallic iron nanoparticles: spectroscopy, electrochemistry and kinetics. *Environ. Sci. Technol.*, 39:1221–1230.
2. French, R.H. 2000. Origins and applications of London dispersion forces and Hamaker constants in ceramics. *J. Am. Ceram. Soc.*, 83:2117–2146.
3. Baker, M.N. 1948. The Quest for Pure Water: The History of Water Purification from the Earliest Records to the Twentieth Century, p. 299–320. New York: The American Water Works Association, Inc.
4. Ackler, H.D., R.H. French, and Y.M. Chiang. 1996. Comparisons of Hamaker constants for ceramic systems with intervening vacuum or water: from force laws and physical properties. *J. Colloid Interface Sci.*, 179:460–469.
5. Rao, C.N. R., B.C. Satishkumar, A. Govindaraj, and M. Nath. 2001. Nanotubes. *Chem. Phys. Chem.*, 2:78–105.
6. Rudalevige, T., A.H. Francis, and R. Zand. 1998. Spectroscopic studies of fullerene aggregates. *J. Phys. Chem. A*, 102:9797–9802.

7. Gillespie, D.T. 1976. A general method for numerical simulating the stochastic time evolution of coupled chemical reactions. *J. Comp. Phys.,* 22:403–434.

8. Orth, W.S. and R.W. Gillham. 1996. Dechlorination of trichloroethene in aqueous solution using FeO. *Environ. Sci. Technol.,* 30:66–71.

9. Langel, W. 2005. Computer simulation of surfaces. In *Handbook of Theoretical and Computational Nanotechnology,* Eds. M. Rieth and W. Schommers, Vol. 1:1–54.

10. Dunphy Guzman, K.A., M.P. Finnegan, and J.F. Banfield. 2006. Influence of surface potential on aggregation and transport of titania nanoparticles. *Environ. Sci. Technol.,* 40:7688–7693.

11. Yamaguchi, T., T. Matsuoka, and S. Koda. 2007. Translational friction and momentum dissipation of a solute in simple liquid studies by generalized Langevin theory for liquids under external field. *J. Molec. Liquid,* 134:1–7.

12. McCash, E.M. 2001. *Surface Chemistry.* New York: Oxford University Press.

13. Tabor, D. and R.H.S. Winterton. 1969. Direct measurement of long range forces between two mica surfaces. *Proc. R. Soc.,* A312:435–450.

14. Chen, K.L. and M. Elimelech. 2007. Influence of humic acid on the aggregation kinetics of fullerene (C60) nanoparticles in monovalent and divalent electrolyte solutions. *J. Colloid Interface Sci.,* 309:126–134.

15. Chen, K.L. and M. Elimelech. 2006. Aggregation and deposition kinetics of fullerene (C60) nanoparticles. *Langmuir,* 22:10994–11001.

16. Labille, J., J. Brant, F. Villieras, M. Pelletier, A. Thill, A. Masioin, M. Wiesner, J. Rose, and J.-Y. Bottero. 2006. Affinity of C60 fullerenes with water. *Fullerenes, Nanotubes and Carbon Nanostructures,* 14:307–314.

17. Lecoanet, H.F., J.Y. Bottero, and M.R Wiesner. 2004. Laboratory assessment of the mobility of nanomaterials in porous media. *Environ. Sci. Technol.,* 38:5164–5169.

18. Hyung, H., J.D. Fortner, J.B. Hughes, and J.H. Kim. 2007. Natural organic matter stabilizes carbon nanotubes in the aqueous phase. *Environ. Sci. Technol.,* 41:179–184.

19. Nowack, B. and T.D. Bucheli. 2007. Occurrence, behavior and effect of nanoparticles in the environment. *J. Environ. Pollut.,* 150:5–22.

20. Fortner, J.D., D.Y. Lyon, C.M. Sayes, et al. 2005. C60 in water: nanocrystal formation and microbial response. *Environ. Sci. Technol.,* 39:4307–4316.

21. Lee, J., J.D. Fortner, J.B. Hughes, and J.H. Kim. 2007. Photochemical production of reactive oxygen species by C60 in the aqueous phase during UV irradiation. *Environ. Sci. Technol.,* 41:2529–2535.

22. Jafvert, C.T. and I. Hua. 2007. Project description: photochemical fate of manufactured carbon nanomaterials in the aquatic environment. http://cfpub.epa.gov/ncer_abstracts/index.cfm/fuseaction/display.abstractDetail/abstract/8404/report/0. (Accessed December 12, 2007)

23. Ferguson, P.L., G.T. Chandler, and W.A. Scrivens. 2004. Project description: chemical and biological behavior of carbon nanotubes in estuarine sedimentary systems. http://cfpub.epa.gov/ncer_abstracts/index.cfm/fuseaction/display.abstractDetail/abstract/7153/report/0. (Accessed December 12, 2007)

24. Ferguson, P., A. DeMarco, R. Templeton, and G. Chandler. 2005. Fate of single-walled carbon nanotubes in the estuarine environment. Presented *at Society for Environmental Toxicology and Chemistry,* Baltimore, MD. 17 November.

25. Weber, W.J. and Q. Huang. 2006. Project description: carbon nanotubes: environmental dispersion states, transport, fate and bioavailability. http://cfpub.epa.gov/ncer_abstracts/index.cfm/fuseaction/display.abstractDetail/abstract/8349/report/0. (Accessed December 12, 2007)

26. Lecoanet, H.F. and M.R. Wiesner. 2004. Velocity effects on fullerene and oxide nanoparticle deposition in porous media. *Environ. Sci. Technol.,* 38:4377–4382.

27. Phenrat, T., N. Saleh, K. Sirk, R.D. Tilton, and G.V. Lowry. 2007. Aggregation and sedimentation of aqueous nanoscale zerovalent iron dispersions. *Environ. Sci. Technol.,* 41:284–290.
28. Elliott, D.W. and W.X. Zhang. 2001. Field assessment of nanoscale bimetallic particles for groundwater treatment. *Environ. Sci. Technol.,* 35:4922–4926.
29. Latif, B. 2006. Nanotechnology for site remediation: fate and transport of nanoparticles in soil and water systems. Prepared for U.S. Environmental Protection Agency, Office of Solid Waste and Emergency Response, Office of Superfund Remediation and Technology Innovation, Technology Innovation and Field Services Division, Washington, D.C. (August).
30. Jin, Y. and J. Xiao. 2006. Project Description: Agglomeration, Retention, and Transport Behavior of Manufactured Nanoparticles in Variably-Saturated Porous Media. http://cfpub.epa.gov/ncer_abstracts/index.cfm/fuseaction/display.abstractDetail/abstract/8348/report/0. (Accessed December 13, 2007)
31. Pennell, K.D., L.M Abriola, and J. Hughes. 2006. 2006 Progress Report: Fate and Transport of Carbon Nanomaterials in Unsaturated and Saturated Soils. http://cfpub.epa.gov/ncer_abstracts/index.cfm/fuseaction/display.abstractDetail/abstract/7834/report/2006. (Accessed December 13, 2007)

7 Treatment of Nanoparticles in Wastewater

Kim M. Henry
AMEC Earth & Environmental

Kathleen Sellers
ARCADIS U.S., Inc.

CONTENTS

Commercial products incorporating nanomaterials eventually reach the end of their usable life. Sunbathers wash sunscreen containing titanium dioxide (TiO_2) nanoparticles from their skin; antimicrobial silver particles drain from washing machines in the rinse cycle; paints and coatings flake; or materials are landfilled. What happens to those nanoparticles at the end of product life? In short, no one knows. Initial attention has focused on the fate of nanoparticles in wastewater treatment. Nanoparticles can enter a municipal wastewater treatment plant as a result of commercial use and discharge. Wastewater discharges from manufacturing processes also can contain nanoparticles. As illustrated by examples in this chapter, however, the discharge and fate of nanomaterials is difficult to quantify.

The same unique properties that make nanomaterials so promising in a wide variety of industrial, medical, and scientific applications may pose challenges with respect to wastewater treatment. In 2004, because the toxicity of nanomaterials and their fate and transport in the environment were not well understood at the time, the British Royal Society and the Royal Academy of Engineering recommended that "factories and research laboratories treat manufactured nanoparticles and nanotubes as if they were hazardous, and seek to reduce or remove them from waste streams" [1]. Although the body of research regarding the toxicity, fate, and transport of nanoparticles has grown [2], literature surveys in 2006 and 2007 indicate that the behavior of nanomaterials during wastewater treatment has not been well studied [3, 4]. An abstract for a research project to evaluate the removal of various types of nanoparticles during wastewater treatment, which was funded by the U.S. EPA's National Center for Environmental Research (NCER) for the period from 2007 to 2010, states: "Today, almost no information is available on the fate of manufactured nanoparticles during biological wastewater treatment" [5].

This chapter discusses the potential for various treatment processes to remove nanoparticles from waste streams. A general description of each process is provided, as well as an evaluation of how particular properties of nanomaterials can reduce or enhance the effectiveness of the process. Research findings are provided where available, or an indication is given as to whether research is ongoing at the time of writing this book. While the primary focus is treatment processes in a typical municipal wastewater treatment plant, many of these processes are used in industrial wastewater treatment. Certain processes also may apply to drinking water treatment and, where relevant, the findings from water treatment research are also discussed.

7.1   MASS BALANCE CONSIDERATIONS

Concerns over the presence of nanoparticles in wastewater streams, which could eventually accumulate in sewage sludge or discharge to the environment in treated wastewater, must be put into context. The concentration of a nanomaterial in wastewater depends primarily on:

- The amount of local production or use of commercial products containing nanomaterials
- Whether the nanomaterials are fixed in a matrix (such as the carbon nanotubes in a tennis racket) or free (such as TiO_2 nanoparticles in sunscreen)
- The amount of the free nanomaterial in the product
- The fraction that is washed down the drain
- The degree of agglomeration or adsorption occurring in aqueous solution that changes the form of the nanoparticle or removes it from solution
- The extent of dilution

No studies have been published of which the authors are aware that attempt to quantify the discharge of nanomaterials into wastewater treatment plants. Given the recent growth of the industry, the wide variety of materials entering the market, and the confidentiality of their formulation, this comes as no surprise. Two case studies illustrate both the potential for nanomaterials to enter wastewater streams and the difficulty in making such an estimate when the details of product manufacture are proprietary. Coincidentally, both examples concern the discharge of silver when washing clothes.

7.1.1 CASE STUDY: SILVERCARE™ WASHING MACHINE

Samsung's SilverCare™ option on several models of washing machine uses silver ions to sanitize laundry. Samsung reportedly spent $10M to develop this technology [6]. The details of the technology are, understandably, proprietary. Company litera-ture describes the technology in several ways. According to one account [6], the sys-tem electrolyzes pure silver into nano-sized silver ions "approximately 75,000 times smaller than a human hair"; assuming that a human hair is approximately 60 to 120 micrometers (μm) wide [4], then the silver nanoparticles would be on the order of 1 nm in diameter. Elsewhere [7], Samsung described their system as follows:

> "[A] grapefruit-sized device alongside the [washer] tub uses electrical currents to nano-shave two silver plates the size of large chewing gum sticks. The resulting positively charged silver atoms — silver ions (Ag^+) — are injected into the tub during the wash cycle."

These two descriptions differ enough to make it unclear whether the silver is released as a true nanoparticle (ca. 1 nm diameter) or as ionic silver. (Silver has an atomic diameter of 0.288 nm and an ionic radius of 0.126 nm [8], and thus silver ions are smaller than the nanoparticle size range of 1 to 100 nm.) Based on the electrolysis process, both may be present. Key and Maas [9] indicate that electrolysis of a silver electrode in deionized water produces colloidal silver containing both metallic silver particles (1 to 25 wt%) and silver ions (75 to 99 wt%). The silver particles observed in colloidal silver generally range in size from 5 to 200 nm; a particle 1 nm in diameter would consist of 31 silver atoms. This information suggests — but certainly does not conclusively prove — that the SilverCare™ washing machine discharges a mixture of silver ions and silver nanoparticles. Silver ions, rather than nanoparticles, may comprise most of the mass.

Samsung has offered several indications of the amount of silver released when washing a load of clothing. Their product literature notes that electrolysis of silver generates up to 400 billion silver ions during each wash cycle [6, 10]. The two chew-ing-gum sized plates of silver reportedly last for 3000 wash cycles [10]. Finally, Samsung reportedly has indicated that using a SilverCare™ washing machine for a year would release 0.05 g silver [11].

With respect to the sanitizing function that this release of silver provides, Sam-sung has indicated that the silver ions "eradicate bacteria and mold from inside the washer" and "stick to the fabric" of clothes being washed to provide antibacterial function for up to 30 days [10]. A Samsung representative stated that "silver nano ions can easily penetrate 'non-membrane cell' [sic] of bacteria or viruses and sup-press their respiration which in turn inhibit [sic] cell growth. On the other hand, Silver Nano is absolutely harmless to the human body" [6].

While Samsung has marketed this antibacterial action as a benefit to customers, some consumers have become concerned about the potential consequences of using

SilverCare™ products. Initial efforts to market the washing machine met with resistance in Germany and the washing machine was taken off the market in Sweden for a brief time due to concerns over the potential toxic effects of discharging silver nanoparticles from the use of these machines to wastewater treatment plants [11, 12]. Chapter 4 discusses regulatory actions in the United States regarding such washing machines.

Attempts to quantify the discharge of silver from using the washing machine — and thus illuminate the potential effects on a municipal wastewater treatment plant — provide a range of answers based on the available data. In addition to the information provided above regarding the mass and potential form of silver released, the following assumptions about wastewater generation were used to complete a conservative mass balance:

- Each wash cycle uses 12.68 gallons of water [13].
- The typical residence generates approximately 70 gallons of wastewater per person per day [14].
- A four-person household does two loads of laundry per day on average.
- All the silver generated in the washing machine enters the sewage.

Further, the authors measured the size of a stick of gum at approximately 0.2 by 1.8 by 7.2 cm, assumed that the density of a silver bar was 10.4 g/cm^3 [8], and conservatively assumed that the entire mass of silver in the two plates would be entirely consumed within the 3000-cycle lifetime.

As a first approximation, the amount of nanosilver particles that could enter a wastewater treatment plant from the use of SilverCare™ in washing clothes could range from 0.001 micrograms per liter (µg/L) to an extreme upper bound concentration of 9 µg/L. The lowest estimate is based on the reported release of 0.05 g silver per year and the assumption that only 25% of the mass would comprise nanoparticles (rather than ions) of silver. The highest estimate is based on complete consumption of the two silver plates during the unit lifetime and the assumption that 75% of the silver was in nanoparticulate form. The actual concentration of nanoparticles would be lower than either of these estimates due to adsorption and agglomeration. Laboratory experiments with solutions of 25-nm and 130-nm silver particles showed that upon vortex mixing, the silver agglomerated into particles ranging up to 16 µm in diameter, well outside the nanoparticle range [15]. Further, the mass balance calculations do not account for dilution by sources of wastewater other than domestic sewage from homes using SilverCare™ washing machines. Dilution from other sources would also decrease the concentration of silver nanoparticles. Thus, the upper bound estimate of 9 µg/L should be regarded as an extreme upper bound.

What effect could this discharge of silver have on the microorganisms in a wastewater treatment plant? As described previously, silver has antimicrobial properties. At the time this book was written, the authors could not identify published benchmarks that enabled them to directly compare the estimated discharge of silver nanoparticles to levels that are either "safe" or "toxic" to microorganisms at a sewage treatment plant. The acute ambient water quality criterion for silver, which was not derived specifically for nanoparticles, is 3.2 µg/L [16]. This concentration is comparable to the upper bound estimate of the discharge of silver nanoparticles into wastewater from using the SilverCare™ system; however, as noted above, that upper bound estimate was quite conservative. As described below, research on the toxicity of silver nanoparticles provides further relevant information.

Rojo et al. [17] assayed the toxicity of colloidal silver nanoparticles in the 5- to 20-nm size range to zebrafish embryos. They tested solutions containing between 1 and 5000 µg/L silver nanoparticles. Their initial tests showed no effect on development or survival of the embryos in the first 2 weeks. Subsequent experiments monitored effects on eight selected genes. At the highest nanosilver concentrations tested, the researchers "found a clear effect on gene expression in most cases." Those concentrations were, however, orders of magnitude higher than the estimated levels of silver nanoparticles in wastewater described above.

Other researchers have worked with mammalian cell lines to test the toxicity of silver nanoparticles. Hussain et al. [18] tested the effect of solutions containing 10 to 50 µg/L silver nanoparticles (15 nm) on PC-12 cells. This neuroendocrine cell line originated from *Rattus norvegicus* (Norwegian rat). The research team observed decreased mitochondrial function in the PC-12 cells upon exposure to the silver nanoparticles. Skebo et al. [15] showed that rat liver cells could internalize silver nanoparticles (25, 80, 130 nm) but that agglomeration of nanoparticles can limit cell penetration. Finally, Braydich-Stolle et al. [19] tested the effects of 15-nm silver nanoparticles on a cell line established from spermatogonia isolated from mice. The nanoparticles reduced mitochondrial function and cell viability at a concentration between 5 and 10 µg/mL (or 5000 and 10,000 µg/L). The researchers estimated the EC50, or the concentration that would provoke a response half-way between the baseline and maximum response, at 8750 µg/L. This level is orders of magnitude higher than the first approximation estimates of silver nanoparticles in wastewater from using the SilverCare™ system.

7.1.2 CASE STUDY: SOCKS WITH NANO SILVER

Several manufacturers market socks impregnated with nanosilver particles as an antibacterial agent. Westerhoff's [20] team at Arizona State University measured the amount of silver that five different brands of socks could release when washed. They simulated washing by placing the socks in deionized water for 24 hours (hr) on an orbital mixer, removing, drying, and then rewashing the socks three times (for a total of four wash cycles). Four of the test socks initially contained silver at 2.0 to 1360 µg/g sock. The fifth sock contained no measurable silver. The amount of silver that leached out of the silver-bearing socks after four simulated wash cycles ranged from 0 to 100%. The concentration of silver in the wash water ranged from less than 1 to 600 µg in 500 mL wash water, or up to 300 µg/L. The research team noted that it was difficult to distinguish between silver ions, silver nanoparticles, and aggregated silver nanoparticles in the wash water.

These initial laboratory results are difficult to extrapolate to the concentration of silver that might result in sewage from washing socks containing silver nanoparticles. As noted above, the typical wash cycle uses more than 12 gallons of water (rather than 500 mL) and runs for much less than 24 hr, suggesting that dilution and a shorter leaching time might result in lower concentrations than were measured in the experiment. The difference in the volume of wash water alone might account for dilution by a factor of 25; additional dilution by other sources of wastewater would reduce the concentration still further. The most difficult variable to quantify would be the number of socks washed per load of laundry (although as any parent would attest, that variable could increase the estimated discharge of silver by at least an order of magnitude over the estimate from washing a single sock).

As these examples show, estimating the discharge of nanomaterials from the use of commercial products is no simple matter. The mass or concentration released to the environment depends on the amount and availability of the material, among other factors, and such proprietary information can be difficult to obtain. The possible effects of exposure can only be inferred from the developing toxicological database. Some research is beginning to produce information on the possible fate of nanomaterials once released; the next section of this chapter describes the fate of nanomaterials in a municipal wastewater treatment plant.

7.2   TREATMENT PROCESSES

Municipal wastewater treatment plants are designed to accelerate the natural processes that remove conventional pollutants, such as solids and biodegradable organic material, from sanitary waste. Treatment processes include:

- Physical treatment, to screen out or grind up large-scale debris, to remove suspended solids by settling or *sedimentation*, and to skim off floating greases
- Biological treatment, to promote degradation or consumption of dissolved organic matter by microorganisms cultivated in *activated sludge* or trickling filters
- Chemical treatment, to remove other constituents by chemical addition, or to destroy pathogenic organisms by *disinfection*
- Advanced treatment, to remove specific constituents of concern by such processes as activated carbon, *membrane separation*, or ion exchange

Similar processes are used in drinking water treatment. *Coagulation*, by the addition of alum and other chemicals, removes suspended solids that cause turbidity and objectionable taste and odors. The *floc* formed during coagulation is removed by sedimentation. *Sand filters* or other porous media such as charcoal subsequently remove smaller particles that remain in suspension. (While more commonly used in water treatment than wastewater treatment, some wastewater treatment systems do incorporate sand filtration.) *Disinfection* removes bacteria or microorganisms [21].

Processes indicated in italic font above are discussed with regard to their potential to remove nanoparticles from waste streams.

7.2.1   SEDIMENTATION

Sedimentation or settling is intended to remove suspended inorganic particles that are 1 μm in size or greater. Because of their size, free non-agglomerated nanoparticles will not be removed during settling, unless by the action of coagulants or flocculants or by the adsorption of the nanoparticles onto large particles [3]. For further discussion of the forces affecting the settling of nanoparticles, see Chapter 6.

7.2.2 COAGULATION AND FLOCCULATION

Coagulation and flocculation are typically used to remove solids in water treatment; certain wastewater treatment applications can include these processes. Coagulation can facilitate the removal of nanomaterials prior to sedimentation or membrane separation [3].

Coagulation refers to the net reduction in electrical repulsive forces at particle surfaces to allow them to agglomerate. In a treatment plant, operators rapidly mix a coagulant (such as aluminum or iron salts, or long-chain polyelectrolytes) into the water to destabilize colloids. Flocculation is the process of aggregating those particles by chemical bridging between particles. After the coagulation step, water is slowly mixed to allow particles to collide and floc to form. Sedimentation removes the floc, or membrane separation can be used to polish the water.

Huang et al. [22] performed jar tests to evaluate the optimal dosage of the coagulant poly-aluminum chlorate (PACl) and the optimum pH required to remove nanoscale silica from chemical mechanical polishing wastewater generated from semiconductor manufacturing. Prior to use, the silica present in the polishing slurry has a uniform particle size of 100 nm. After the polishing process, the colloidal silica particles present in the wastewater range in size from 78 to 205 nm and, without pretreatment, can penetrate and clog the microfiltration membrane. The researchers found that supernatant from the jar tests had the lowest turbidity when the pH was around 6 and the concentration of PACl was greater than 10 mg/L. At pH 6, the PACl acts to neutralize the negatively charged silica and to destabilize the colloidal particles. Supernatant representative of the range of optimal conditions identified in the settleability tests was then subjected to filterability testing by measuring the time to pass 50 mL of the supernatant through the microfiltration membrane. This testing confirmed that a pH of 6 and a PACl concentration of 30 mg/L produced the shortest filtration time. The coagulation enlarged the particle size such that nearly all the particles were greater than 4000 nm in diameter. Although subsequent microfiltration through a 500-nm membrane removed approximately 95% of the silica, silica still remained in the treated wastewater at a concentration of 44 mg/L [22].

Kvinnesland and Odegaard [23] studied the effect of different polymers on the coagulation and flocculation of humic substances present in water primarily as nanoparticles less than 100 nm in size. For the purposes of their study, they defined coagulation as the process by which the nanoparticles formed aggregates that could be removed by a 100-nm filter, and flocculation as the process by which the particles further agglomerated for removal by an 11,000-nm filter. The researchers found that the five different polymers achieved the same maximum removal of nanoparticles via coagulation (approximately 95% removal). The coagulation was achieved by the addition of cationic charge regardless of the type of polymer applied. Removal of the humic substances by flocculation varied according to the charge density of the different polymers [23].

In a project funded by NCER for the period from 2004 to 2007, Westerhoff et al. [24, 25] are researching the fate, transformation, and toxicity of manufactured nanomaterials in drinking water. As part of their research, they have conducted jar tests of coagulation, flocculation, sedimentation, and filtration to evaluate the removal of metal oxide nanoparticles during typical drinking water treatment processes. The metal oxide nanoparticles are present in solution as stable aggregates that range in size from 500 to 10,000 nm [24]. Metal coagulants (alum) and salt (magnesium chloride) were added to solutions of commercial metal oxide nanoparticles, lab-synthesized hematite nanoparticles, and cadmium quantum dots. According to a paper presented at the NSTI-Nanotech 2007 Conference [25], "removal of nanomaterials by coagulation, flocculation and sedimentation processes was relatively difficult." More than 20% of the commercial metal oxide and the laboratory-synthesized hematite nanoparticles remained in the water following these processes. For all the nanoparticles tested, microfiltration through a 0.45-μm filter following sedimentation removed additional nanoparticles. However, 5 to 10% of the initial concentration of particles remained after completion of the simulated drinking water treatment process [25].

The presence of other constituents in the water can affect the coagulation and flocculation of nanoparticles. In a presentation to the National Institute of Environmental Health Sciences, Westerhoff suggests that dissolved organic matter (DOM) present in water may stabilize nanoparticles by inhibiting the formation of aggregates. The DOM thus affects the removal of nanoparticles during sedimentation and filtration [20]. For example, Fortner et al. [26] have conducted research on the factors that affect the formation of nano-C60, the water-stable aggregate that forms when fullerenes (C60) come in contact with water. Their research shows that the pH of the water affects the particle size of the nano-C60, and the ionic strength affects the stability of the nano-C60 in solution [26].

Similarly, multi-walled carbon nanotubes are hydrophobic and would be expected to aggregate and settle out in water. However, researchers at the Georgia Institute of Technology have observed that multi-walled nanotubes adsorb to organic material that occurs naturally in river water, forming a suspension that persisted for the month-long period of observation. The natural organic matter appeared to be a better stabilizing agent than sodium dodecyl sulfate, a surfactant often applied in industrial processes to stabilize carbon nanotubes [27]. This type of interaction of nanoparticles with constituents in natural waters would likely affect their removal.

7.2.3 ACTIVATED SLUDGE

Some nanoparticles can be removed by adsorption to activated sludge [3]. A research project funded by NCER for the period from 2007 to 2010 will address the fate of manufactured nanoparticles during biological wastewater treatment. The investigators (Westerhoff, Alford, and Rittman of Arizona State University) indicate that the objective of their research is to quantify the removal of four types of nanoparticles (metal-oxide, quantum dots, C60 fullerenes, and carbon nanotubes) during wastewater treatment. Batch adsorption experiments will be performed using whole biosolids, cellular biomass only, and extracellular polymeric substances from biological reactors and full-scale wastewater treatment reactors. Nanoparticles also will be added to laboratory-scale bioreactors to quantify biotransformations to the nanoparticles and toxicity to the microorganisms. Electron microscopy imaging will be used to evaluate the interactions between the nanoparticles and the biosolids [5].

No NCER progress reports were available for the research of Westerhoff, Alford, and Rittman at the time of writing this book. However, the investigators hypothesize in their research abstract that "dense bacterial populations at wastewater treatment plants should effectively remove nanoparticles from sewage, concentrate nanoparticles in biosolids, and/or possibly biotransform nanoparticles. The relatively low nanoparticle concentrations in sewage should have negligible impact on the wastewater treatment plant's biological activity or performance" [5]. Preliminary results [20] hint at the possible behavior of C60 fullerenes in sewage treatment. In initial tests, the research team mixed a solution of C60 aggregates and biomass in water, then filtered the solids and measured C60 levels to determine the amount sorbed to biosolids. These results were incorporated into mass balance modeling that simulated the operation of a wastewater treatment plant at steady state. The results indicated that 22% of C60 would adsorb to biosolids and the remainder would be discharged in the effluent. Westerhoff [20] noted that the model estimates must be validated with laboratory and field measurements.

Ivanov et al. [28] conducted research to evaluate whether microbial granules present in a biofilm could remove nano- and micro-particles from wastewater and whether calcium enrichment, which is typically applied to wastewater with high organic loading, could enhance the removal of small particles. Calcium ions enhance the formation of microbial aggregates by decreasing the negative surface charge of the cells. Therefore, particle removal by microbial granules was evaluated for different calcium concentrations. Two laboratory-scale sequencing batch reactors, one with no calcium supplement and the other with a calcium concentration of 100 mg/L, were inoculated with aerobic sludge and operated in parallel. The influent consisted of synthetic wastewater. Aerobic granules from the reactors were incubated with particle suspensions of different sizes: 100-nm fluorescent microspheres, 420-nm fluorescent microspheres, and stained cells of *Escherichia coli*. Researchers used a confocal laser scanning microscope, a flow cytometer, and a fluorescence spectrometer to measure the rate of particle removal and the accumulation of particles in the microbial granules. The results showed that the addition of calcium did not enhance the removal of microspheres from the wastewater. Microspheres were adsorbed to the surface of the granules but the depth of penetration did not vary with the calcium concentration, as it did for the *E. coli* cells [28]. Ivanov et al. concluded that the behavior of inorganic nanoparticles in aerobic wastewater treatment is different from the behavior of biological cells.

Researchers have shown that at certain concentrations, some nanoparticles may be toxic to bacteria. For example, Fortner et al. [26] have shown that nano-C60 inhibits the growth of bacterial cultures at concentrations of 0.4 mg/L or more and decreases aerobic respiration rates at 4 mg/L. Other research supports the antibacterial activity of nano-C60 water suspensions, indicating that suspensions formed by four different processes exhibited minimum inhibitory concentrations ranging from 0.1 to 1.0 mg/L [29]. As noted previously, silver also can have antimicrobial activity.

7.2.4 SAND FILTERS

Brownian diffusion is the dominant mechanism governing the transport of nanoparticles through the granular filter. As they pass through the filter, nanoparticles are removed from the fluid stream by several processes, including:

1. Brownian diffusion causes the nanoparticles to agglomerate into larger particles or to agglomerate with the filter grains.
2. Nanoparticles are immobilized by gravitational sedimentation because their density is higher than that of the filter medium, or the flow velocity is reduced within the filter bed.
3. Nanoparticles are intercepted by physical contact with the filter medium [30]. Attachment of particles to the filter medium is affected by a variety of forces, described by the term "attachment efficiency," as discussed further below [31].

The attachment efficiency (α) is the ratio of the rate of particle deposition to the rate of particle collisions with the filter medium [31]. This parameter is governed by various phenomena, including van der Waals forces, the forces of solvency, and electrostatic repulsive forces (see Chapter 6). When α is less than unity, conditions are not conducive to particle attachment. When α equals unity, no barriers to particle attachment exist. When α is greater than unity, particles may be attracted to the surface of the filter medium over small distances. However, for very small nanoparticles less than 2 nm in size, the relative effects of the forces governing the parameter α can be unpredictable and different from those of larger particles. If smaller nanoparticles aggregate to form colloidal material, as has been observed for C60 fullerenes and some other particles, the behavior of the material within a granular filter will differ from the response predicted based on the size of the original manufactured particle. Therefore, researchers have concluded that direct measurement of the mobility of nanoparticles is currently the most accurate means by which to quantify their behavior in porous media [32].

Nanoparticle mobility within a porous medium is a function not only of size, but also of surface chemistry [32]. Lecoanet, Bottero, and Wiesner [30] conducted laboratory experiments to quantify the mobility of eight different manufactured nanomaterials in a porous medium of glass beads, which the researchers indicated would be representative of a water treatment plant filter or a sandy groundwater aquifer. Their results indicated that different forms of nanoparticles with the same composition have different mobilities. For example, of the carbon-based particles tested, single-walled nanotubes and fullerols (hydroxylated C60) passed through the porous medium more rapidly than the colloidal aggregate form of C60 known as nano-C60. The solubilized forms of the particles are more mobile than the suspended form [33].

Conditions in the waste stream, such as pH and ionic strength, will also affect the behavior of nanoparticles in water and the attachment efficiency of nanoparticles passing through a filter medium [31]. As noted above, Fortner et al. [26] observed that the pH and ionic strength of water affect, respectively, the particle size and stability of the nano-C60 in solution.

Finally, surface coatings applied to manufactured nanoparticles also will affect their mobility in porous media. Typical surface coatings include polymers, polyelectrolytes, and surfactants, and are often applied with the intention of improving the delivery or mobility of the nanoparticles. Because these coatings can affect the surface charge of the nanoparticles or stabilize the particles against aggregation, they may reduce the ability of the filter medium to remove the nanomaterials from the waste stream [32].

7.2.5 Membrane Separation

Membrane separation is the general process in which contaminants are removed from a fluid as it passes through a microporous membrane. Specific membrane processes are distinguished by the size of the pores or the size of the particles retained by the membrane, as follows [34]:

*   Microfiltration (MF): 100 to 10,000 nm
*   Ultrafiltration (UF): 1 to 100 nm
*   Nanofiltration (NF): 0.1 to 1 nm
*   Reverse osmosis (RO): less than 0.1 nm

Microfiltration of individual or agglomerate nanoparticles of 100 nm or more in size can result in fouling the membrane. Particles less than 100 nm in size can pass through the membrane.

The smaller particles must be pretreated by coagulation prior to the microfiltration (see discussion above), or treated by other means [3]. Figure 7.1 shows the ranges over which these various forms of filtration can generally be effective.

7.2.6 Disinfection

A research project funded by the NCER for the period from 2005 to 2008 focuses on the fate and transformation of C60 nanoparticles in water treatment processes. In the 2006 progress report, the investigators, Kim and Hughes, documented the results of applying dissolved ozone, a common disinfectant reagent, to a suspension containing the aggregate nano-C60. The products of this treatment were highly oxidized, soluble fullerenes [38], suggesting that disinfection has the effect of rendering a stable aggregate more soluble and thus potentially more mobile. Future research activities will include applying ultraviolet radiation and chlorine to water containing nano-C60 [38].

7.3   SUMMARY

At this early point in the nanotechnology revolution, we know little about the fate of nanomaterials at the end of useful product life. The amount of nanomaterial released to the environment may be limited by the relatively low concentrations of free nanomaterials in many products; the mobility of those nanomaterials, once released, may be limited by agglomeration and adsorption. However, few relevant data now

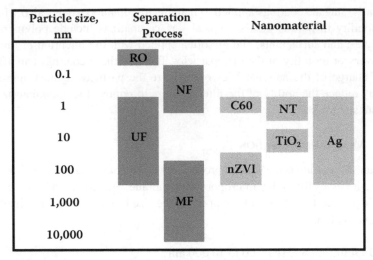

Notes: RO = reverse osmosis, NF = nanofiltration, UF = ultrafiltration,
MF = microfiltration, C60 = fullerene, nZVI = zero valent iron,
NT = nanotube, TiO$_2$ = titanium dioxide, Ag = silver

FIGURE 7.1 Applicability of membrane separation processes relative to nanoparticle size. (From References 9, 25, 33, and 35–37.)

exist, and manufacturers' need to protect confidential business information can limit access to relevant data.

Many initial concerns about the end-of-life fate of nanomaterials focus on wastewater treatment. Initial research shows some potential for removal in various unit processes. The extent of that removal, and the potential toxic effects of those nanomaterials, vary substantially between materials. Particle size, concentration, and surface properties, as well as the other characteristics of the wastewater, can affect removal.

REFERENCES

1. The Royal Society and the Royal Academy of Engineering. 2004. *Nanoscience and Nanotechnologies: Possible Adverse Health, Environmental and Safety Impacts*, Chapter 5. July. http://www.nanotec.org.uk/report/chapter5.pdf. (Accessed September 18, 2007)
2. Dunphy Guzman, K.A., M.R. Taylor, and J.F. Banfield. 2006. Environmental risks of nanotechnology: National Nanotechnology Initiative funding, 2000–2004. *Environ. Sci. Technol.*, 40:1401–1407.
3. Ganesh, R. and L.Y.C. Leong. 2006. Potential impacts of nanoparticles on water reclamation. http://www.deq.state.va.us/wastewater/documents/PotentialImpacesofNano particlesonWaterReclamation.pdf. (Accessed September 24, 2007)
4. U.S. Environmental Protection Agency. 2007. *USEPA Nanotechnology White Paper.* EPA 100/B-07/001.
5. Westerhoff, P., T. Alford, and B. Rittman. 2007. EPA grant description: Biological fate and electron microscopy detection of nanoparticles during wastewater treatment. http://cfpub.epa.gov/ncer_abstracts/index.cfm/fuseaction/display.abstract-Detail/abstract/8402/report/0. (Accessed September 23, 2007)

6. Samsung. 2005. Press Release: Samsung Silver Nano Health System™ Gives Free Play to Its 'Silver' Magic. 29 March.http://www.samsung.com/he/presscenter/pressrelease/pressrelease_20050329_0000109066.asp (accessed September 29, 2007).

7. Samsung. 2006. Press Release: Samsung Laundry Featuring Silver Care™ Technology. 13 February. http://www.samsung.com/us/aboutsamsung/news/newsRead.do?newstype=productnew&newsctgry=consumerproduct&news_seq=3071. (Accessed September 29, 2007)

8. Etris, A.F. 2001. Silver and silver alloys. In *Kirk-Othmer Encyclopedia of Chemical Technology*, 4:761–803. John Wiley & Sons, Inc.

9. Key, F.S. and G. Maas. 2001. Ions, atoms and charged particles. http://www.silver-colloids.com/papers/IonsAtoms&ChargedParticles. (Accessed October 6, 2007)

10. Samsung. FAQs: What is SilverCare™? http://ars.samsung.com/customer/usa/jsp/faqs/print.jsp?AT_ID=73145. (Accessed September 21, 2007)

11. Nanowerk News. 2006. Concerns about nanotechnology washing machine. 16 November. http://www.nanowerk.com/news/newsid=1037.php. (Accessed September 21, 2007)

12. Senjen, R. 2007. Nanosilver — a threat to soil, water and human health?. Friends of the Earth Australia. March. http://www.nano.foe.org.au. (Accessed August 2007)

13. Samsung. FAQs: How Much Water Does My Samsung Washer Use Per Cycle?. http://ars.samsung.com/customer/usa/jsp/faqs/print.jsp?AT_ID=70294. (Accessed September 21, 2007)

14. Tchobanolglous, G. and F.L. Burton. 1991. *Wastewater Engineering: Treatment, Disposal and Reuse*, 3rd edition, Chapter 4, p. 27. Metcalf & Eddy, Inc. New York: McGraw-Hill Publishing Company.

15. Skebo, J.E., C.M. Grabinsky, A.M. Schrand, J.J. Schlager, and S.M. Hussain. 2007. Assessment of Metal Nanoparticle Agglomeration, Uptake, and Interaction using High-Illuminating System. *Int. J. Toxicol.*, 26:135–141.

16. U.S. Environmental Protection Agency. Current National Recommended Water Quality Criteria. http://earth1.epa.gov/waterscience/criteria/wqcriteria.html. (Accessed September 30, 2007)

17. Rojo, I., M. Uriarte, I. Bustero, A. Egizabal, M.A. Pardo, and O. Martinez de Ilarduya. 2007. Toxicogenomics study of nanomaterials on the model organism Zebrafish. NSTI-Nanotech 2007, www.nsti.org. ISBN 1420061836 (Vol. 2).

18. Hussain, S.M., A.K. Javorina, A.M. Schrand, H.M. Duhart, S.F. Ali, and J.J. Schlager. 2006. The interaction of manganese nanoparticles with PC-12 cells induces dopamine depletion. *Toxicolog. Sci.*, 92(2):456–463.

19. Braydich-Stolle, L., S. Hussain, J.J. Schlager, and M.C. Hofmann. 2005. *In vitro* cytotocicity of nanoparticles in mammalian germline stem cells. *Toxicolog. Sci*, 88(2):412–419.

20. Westerhoff, P. 2007. Nanoparticle interactions during wastewater and water treatment. Presented at *Nanotechnology — Applications and Implications for Superfund, Session 6: Nanotechnology — Fate and Transport of Engineered Nanomaterials*. National Institute of Environmental Health Sciences, Superfund Basic Research Program, U.S. Environmental Protection Agency. 16 August.

21. U.S. Environmental Protection Agency. 2000. The History of Drinking Water Treatment. EPA-816-F-00-006 (February).

22. Huang, C., W. Jiang, and C. Chen. 2004. Nano silica removal from IC wastewater by pre-coagulation and microfiltration. *Water Sci. Technol.*, 50(12):133–138.

23. Kvinnesland, T. and H. Odegaard. 2004. The effects of polymer characteristics in humic substances removal by cationic polymer coagulation. *Water Sci. Technol.*, 50(12):185–191.

24. Westerhoff, P., D. Capco, Y. Chen, and J.C. Crittenden. 2005. 2005 Progress Report: Fate, Transformation and Toxicity of Manufactured Nanomaterials in Drinking Water. http://cfpub1.epa.gov/ncer_abstracts/index.cfm/fuseaction/display.abstractDetail/abstract/7387/report/2005. (Accessed September 21, 2007)
25. Zhang, Y., B.A. Koeneman, Y. Chen, P. Westerhoff, D.G. Capco, and J. Crittenden. 2007. Fate, transport and toxicity of nanomaterials in drinking water. *Technical Proceedings of the 2007 Nanotechnology Conference and Trade Show*, ISBN 1420061836, 2:678–680.
26. Fortner, J.D., D.Y. Lyon, C.M. Sayes, et al. 2005. C60 in water: Nanocrystal formation and microbial response. *Environ. Sci. Technol.*, 39:4307–4316.
27. Hyung, H., J.D. Fortner, J.B. Hughes, and J.H. Kim. 2007. Natural organic matter stabilizes carbon nanotubes in the aqueous phase. *Environ. Sci. Technol.*, 41:179–184.
28. Ivanov, V., J.H. Tay, S.T.L. Tay, and H.L. Jiang. 2004. Removal of micro-particles by microbial granules used for aerobic wastewater treatment. *Water Sci. Technol.*, 50(12):147–154.
29. Lyon, D.Y., L.K. Adams, J.C. Faulkner, and P.J.J. Alvarez. Antibacterial activity of fullerene water suspensions: Effects of preparation method and particle size. *Environ. Sci. Technol.*, 40:4360–4366.
30. Tufenkji, N. and M. Elimelech. 2004. Correlation equation for predicting single-collector efficiency in physiochemical filtration in saturated porous media. *Environ. Sci. Technol.*, 38:529–536.
31. Lecoanet, H.F. and M.R. Wiesner. 2004. Velocity effects on fullerene and oxide nanoparticle deposition in porous media. *Environ. Sci. Technol.*, 38:4377–4382.
32. Wiesner, M.R., G.V. Lowry, P. Alvarez, D. Dionysiou, and P. Biswas. 2006. Assessing the risks of manufactured nanomaterials. *Environ. Sci. Technol.*, 40:4336–4345.
33. Lecoanet, H.F., J.Y. Bottero, and M.R Wiesner. 2004. Laboratory assessment of the mobility of nanomaterials in porous media. *Environ. Sci. Technol.*, 38:5164–5169.
34. Tetra Pak Processing Systems. 2003. *Dairy Processing Handbook*. Lund, Sweden. http://www.egr.msu.edu/~steffe/handbook/fig641.html. (Accessed September 30, 2007)
35. Taylor, R. 2002. Fullerenes. In *Kirk-Othmer Encyclopedia of Chemical Technology*, 12:228–258. New York: John Wiley & Sons, Inc.
36. Nurmi, J.T., P.G. Tratynek, V. Sarathy, et al. 2005. Characterization and properties of metallic iron nanoparticles: Spectroscopy, electrochemistry, and kinetics. *Environ. Sci. Technol.*, 39:1221–1230. Published online December 16, 2004.
37. Particle Size Distribution Study of Polymetallix™ nZVI. Not Dated. http://www.polymetallix.com/PDF/Particle%20Size%20Distribution%20Study%20of%20Polymetallix%20Nanoscale%20Iron.PDF. (Accessed August 7, 2007)
38. Kim, J.H. and J. Hughes. 2006. 2006 Progress Report: Fate and transformation of C60 nanoparticles in water treatment processes. http://cfpub.epa.gov/ncer_abstracts/index.cfm/fuseaction/display.abstractDetail/abstract/7725/report/2006. (Accessed September 24, 2007).

8 The Potential Ecological Hazard of Nanomaterials

Stephen R. Clough
Haley & Aldrich

CONTENTS

Puzzles eventually have answers; mysteries, however, cannot. Unknowns or uncertainties preclude a definitive answer to a mystery [1]. A mystery "can only be framed, by identifying the critical factors and applying some sense of how they have interacted in the past and might interact in the future. A mystery is an attempt to define ambiguities" [1]. In its infancy, nanotechnology can seem mysterious to both the layperson and the scientist. Science now enables us to construct nanomaterials but, paradoxically, some generally accepted scientific principles do not appear to apply to their inherent biological activity. For example, a substance like gold that is physiologically inert at the microscale has been shown to have biological activity at the nanoscale [2]. This change, in effect, can result from the fact that a particle that is less than 100 nanometers (nm) in size can behave more according to the laws of quantum physics than Newtonian physics. As the science emerges, the mysteries of nanomaterials will become puzzles that will be solved. The scientific paradigms for nanotechnology may take much longer to decipher because conventional scientific methodologies, instrumentation, or principles may not apply in some of the upcoming studies. Many fear that regulations put into place to protect both the workplace and the environment will be too little, too late.

This chapter discusses one of the mysteries surrounding nanotechnology and presents data that scientists will ultimately use to solve the puzzle. It faces the question:

> "If a nanomaterial were to be released into the general environment, would it pose a significant risk to ecological organisms such as fish or wildlife?"

The answer begins with some background information on how toxicologists assess impacts to fish and wildlife, referred to in ecological assessments as "ecological receptors."

8.1 UNDERLYING PRINCIPLES OF ECOLOGICAL EXPOSURE, EFFECTS, AND "RISK"

This section provides a brief primer on ecological risk assessment, to provide the reader with the context for discussing the potential hazards of nanomaterials.

8.1.1 TERRESTRIAL VS. AQUATIC ECOSYSTEMS

Because of obvious differences in habitat, ecotoxicology comprises two main categories of environmental assessment: (1) terrestrial and (2) aquatic. The former category addresses the impacts of chemicals released into the environment on terrestrial species. Examples include invertebrates such as earthworms, bees, beetles, and grubs; birds, including doves, quail, robins, and hawks; reptiles, such as lizards and snakes; and mammals, such as shrews, mice, foxes, or bears. The latter category includes aquatic species, such as phytoplankton (e.g., single or multicellular algae), zooplankton (e.g., rotifers, cladocercans, paramecia), benthic invertebrates and insect larvae (e.g., mayflies, caddisflies, stoneflies) and fish (e.g., embryos, fry, juveniles, or adults). Of course, some animals — for example, amphibians such as frogs, toads, and salamanders — may spend portions of their life cycle in both the aquatic and terrestrial environment. Organisms in a third category, semiaquatic receptors, strongly depend on waterbodies for food or sustenance. These semiaquatic receptors include fish-eating birds (e.g., kingfisher, heron, osprey, and eagle) or mammals whose habitat is primarily aquatic (e.g., beaver, muskrat, and otter).

With the possible exception of some deserts, these different types of habitat are not mutually exclusive. The forces of the water cycle will strongly affect both the fate and the transport of contaminants within a terrestrial ecosystem. In addition, animal activity can affect markedly the landscape of a terrestrial ecosystem. The leg-trapping of beavers, for example, was once an accepted method in the United States to obtain their thick pelts. Many states, however, now view these traps as inhumane and have banned their use. Consequently, their populations are back on the rise and, as a result, their natural impoundments are transforming once-dry forest land into large, productive wetlands.

Because of the limited data available regarding the effects of nanomaterials on ecological receptors in the wild, this chapter first examines the underlying principles that must be in place for there to be a valid supposition that nanomaterials may eventually pose a risk to any terrestrial, aquatic, or semiaquatic organisms/receptors.

8.1.2 RISK AND HAZARD

Risk is generally defined as the probability that a hazard will occur in a given time and space. It is virtually impossible to determine the probability that a chemical may pose a risk to an organism, population, or community in the wild. Thus, the term "ecological risk" is something of a misnomer. The term "hazard," which is the likelihood that an adverse event can take place, better expresses the degree of harm to an ecological receptor. However, these terms often are used interchangeably.

Risk (or hazard) is a function of toxicity and exposure. Unless an ecological receptor is exposed to a chemical or nanomaterial, there can be no risk or hazard. If exposure is great enough, substances that have a low inherent toxicity can still result in a toxic response. Paracelsus, known as the Father of Modern Toxicology, stated that "[a]ll substances are poisons; there is none which is not a poison. The right dose differentiates a poison and a remedy." Thus, if enough of a substance of known (but low) toxicity is ingested, a hazard may exist. Although table sugar is classified as virtually non-toxic, eating too much cake or candy will result in nausea and/or vomiting, a toxic response elicited by the over-consumption of sugar.

The potential for harm also depends on the duration of exposure. Short-, medium-, and long-term contact with the material in question are referred to, respectively, as acute (single dose), subchronic (multiple exposures over 2 to 3 months), and chronic (greater than 3 months to a lifetime) exposures. Over time, some animals can become tolerant to some materials, or cross-tolerant to similar materials. A good example is the highly toxic metal cadmium. An acute exposure of an organism to the metal will impart tolerance or resistance to subsequent exposures due to the induction of metal binding proteins by various tissues.

8.1.3 TOXICITY

Ecological hazard assessments can focus on individuals or populations. Individual organisms can be exposed to nanomaterials via inhalation, dermal contact, and ingestion. Exposure pathways historically have been framed in the context of food webs that embody many different types of autotrophic and heterotrophic interactions. Persistent, bioaccumulative, and/or toxic substances (PBTs) will bioconcentrate, bioaccumulate, and/or biomagnify in a food web.

Scientists generally divide the evidence of ecological harm into two classes of effects criteria: (1) Assessment Endpoints and (2) Measurement Endpoints. They generally ascribe Assessment Endpoints to a less-tangible (or more subjective) value, such as "Will Chemical X, if released into the environment at Concentration Y, have an adverse effect on the population of predatory fish?" A Measurement Endpoint is a more specific, objective measurement at the individual or community level that supports the evaluation of the Assessment Endpoint, such as: "What is the Concentration Y of Chemical X that will adversely affect 20% of a known population of rainbow trout in the laboratory?"

The main endpoint for measuring ecological toxicology is the LD50, or the lethal dose required to kill 50% of the organisms under controlled laboratory testing conditions. For aquatic organisms, the LC50 and EC50 (or the respective lethal concentration and effect concentration required to kill or affect 50% of the organisms) are the more appropriate terms used for a toxicity endpoint. When dose is plotted versus response, the slope of the curve is a general indication of the potency of the toxicant: the steeper the slope, the more potent the toxicant relative to chemicals of a similar class.

One can generalize about how these criteria will reflect the relative toxicity of a substance based on its structure and the principle that like dissolves like. Because cell membranes primarily comprise a lipid bilayer, lipophilic or fat-loving substances are, as a general rule of thumb, more toxic than hydrophilic or water-loving (soluble) substances. Lipophilic substances are more easily absorbed by inhalation, ingestion, or dermal exposure, and tend to have a longer half-life (i.e., the time required to reduce the body burden of a toxicant by one-half, either by metabolism or excretion), while water-soluble substances are more easily metabolized by the liver and/or excreted by the kidney and thus tend to have a shorter residence time in the body.

In the field of inhalation toxicology, foreign matter is generally categorized as gas, vapor, or particulate (or fibrous) matter. The latter can affect physically the elasticity of the lung. Examples include silicosis in concrete and quarry workers, asbestosis in shipyard workers, and pneumoconiosis in coal miners. Nanoparticles would be classified as particulate matter, but because these particulates are so extraordinarily small, they fall in a toxicological gray area. Some comprise potentially toxic elements that, if dissociated or dissolved, may cause adverse effects inside a cell. Therefore, they may cause adverse extracellular physical effects similar to those caused by larger fibers such as asbestos or fiberglass insulation, or may be actively or passively internalized by cells and cause toxic effects by interfering with cellular processes. Data from a battery of both *in vitro* and *in vivo* bioassays may be needed to reveal to the investigator the inherent toxicity of the various elements and compounds that comprise nanomaterials (for some of which there are little to no toxicological data). The difficulty will lie in separating whether an adverse effect reflects a physical effect induced by the nanomaterial or a direct toxic effect resulting from the composition of the material itself.

For example, carbon black, a common nanomaterial in commercial use for decades, is considered biologically inert. Although it may remain in the body in a sequestered form, it is expected to have a low inherent toxicity [3]. In contrast, a unique nanomaterial constructed from one (or more) elements may be inherently toxic. Consider cadmium, a highly toxic metal used to make quantum dot alloys of cadmium selenide or cadmium telluride. Toxic effects on the reproductive system or the nervous system are of particular concern. The response of these systems, in general, will take a longer time to unravel than other biological endpoints, because the endpoints take a long time to achieve, are expensive to characterize, or the results are characteristically subtle, requiring innovative and/or very sensitive testing methodologies.

The natural physiological variability within a population means that individuals may react differently upon exposure. The reasons given for this variability often are physiological, such as internal genetic differences, or environmental. The gender of an animal, the species, or its age can make a very significant difference in the response following exposure to a chemical or nanomaterial. Younger animals are generally more susceptible to toxicants than older animals, partly due to the fact that they weigh less and therefore, pound for pound, will get a larger dose than would an adult animal. Similarly, there are some strains of mice that are very resistant to the toxic effects of heavy metals, whereas other strains are overly sensitive. The results of these variations in sensitivity can be observed in the classic dose/response curve, which is typically an S-shaped function. Plotted on a graph, with the dose on the x-axis and the percent of organisms affected on the y-axis, the cause of the inflections in the S-shaped curve are due to the presence of sensitive individuals in the low dose ranges and tolerant individuals in the high dose ranges.

8.1.4 EXPOSURE

A complete exposure pathway must exist for an animal to be affected by a chemical or nanomaterial. This means that a mechanism must exist to transfer the compound or nanomaterial in question from the *source* in air, water, soil, or sediment to the *receptor* organism in question. *Without exposure, there can be no risk.* Therefore, and this is a critical factor as nanotechnology evolves, as long as nanomaterials are properly handled and/or contained, risk and/or hazard(s) will be negligible.

Scientists use the term "fate and transport" to refer to processes that affect a substance as it travels from the source to a potential receptor. As described in Chapter 6, various processes can change the nature and concentration of a nanomaterial, which, in turn, can change its potential to induce toxicity.

Partitioning from one phase of media to another is an extremely important phenomenon that can affect the properties (and often the quantities) of a nanomaterial within an environmental medium. Partitioning typically is expressed in terms of a ratio or partition coefficient (e.g., water-to-sediment, soil-to-water, water-to-air, water-to-biota, etc.). For example, a bioconcentration factor (BCF) is the ratio of the concentration of a substance in fish tissue to the concentration in a waterbody.

Weathering, which includes the variety of chemical reactions and physical attenuation processes that occur after a chemical is released into the environment, will generally decrease exposure, bioavailability, and/or toxicity. The exceptions to this are compounds or materials that resist degradation, such as mercurials or arsenicals, some types of commercial pesticides, polychlorinated dioxins and furans, and polychlorinated biphenyls, to name just a few examples.

Another important underlying principle in ecological toxicology is the difference between exposure and dose. An *exposure* is the sum total of a compound or nanomaterial that reaches an ecological receptor, but the *dose* is a smaller percentage of the total material that actually enters the body. Bioaccessibility and bioavailability also come into play. The bioaccessible fraction of a substance like a nanomaterial would be the amount of material that can be presented to a tissue or organ for uptake. For example, if a nanomaterial agglomerates, organisms cannot access the inner portion of an intact clump. The outside (exposed) portion of the clumped material may be able to react with receptors on a cell surface or penetrate a cell membrane, and thus would be bioavailable.

Although the degree of risk or hazard that a nanomaterial may pose to an animal is clearly a function of both the degree of exposure and the inherent toxicity of the material, defining the latter two parameters can be quite complex. In bioassays, some researchers will hold the exposure or dose at a steady concentration and then evaluate the effects of the material over time, while others will vary the exposure or dose and stop the experiment or study after a specified time period. The latter generally is preferred as demonstrating dose dependence, a key principle in the science of toxicology. Because nanomaterials can be unique compounds, many of which will be water insoluble and therefore difficult to find a dosing vehicle for, the science of toxicology may have to adapt new and innovative methods for testing many of these distinctive materials as they come into the marketplace.

8.2 FACTORS THAT CAN AFFECT THE TOXICOLOGY OF NANOMATERIALS

Will traditional toxicology testing protocols allow for the proper evaluation of the hazard of a nanomaterial? The answer depends on toxicity and exposure. This section describes the factors that can affect the toxicology of nanomaterials. Sections 8.3 and 8.4 present the results of laboratory studies to date.

8.2.1 TOXICITY OF NANOMATERIALS

Toxicity depends, in part, on particle size, shape, and chemical composition. As discussed previously, a nanomaterial is defined as a substance that measures less than 100 nanometers (nm) in any one of three dimensions. Relatively speaking, that is 100 to 1000 times smaller than most living cells [4]. For another perspective, the size of nanomaterials falls in between the wavelength range of ultraviolet light (450 to 10 nm) and x-rays (<10 nm). Nanomaterials, therefore, are difficult to observe or to detect in the laboratory [5]. As particles get smaller, the surface-to-volume ratio increases dramatically. This large amount of area presents many surfaces that can interact with, and possibly interrupt, normal cellular physiological mechanisms. For example, titanium dioxide (TiO_2) is a relatively inert substance at the microscale, but nanoscale TiO_2 has been shown to produce reactive oxygen species (ROS) with consequent potential for cellular damage in both prokaryotic and eukaryotic cell cultures [6–8].

Size and shape also determine where a material might end up in the body. Upon autopsy, a normal individual's lung will show a pepper-like coloration, both at the surface and upon incision. This coloration results from a lifetime's accumulation of both natural and anthropogenic dusts and soots. The reticulo-endothelial system (or the RES, comprising macrophages, white blood cells, and lymph nodes) sequesters most particulates, making the material unavailable to the rest of the body. Too much exposure, however, will overwhelm the RES and the lung will become fibrotic, calcified, or emphysematous, losing its elasticity and eventually resulting in lung disease. Nanomaterials may pose the greatest risk to the lung because they can be transported like a gas and reach the deepest portion of the lungs, the alveoli. The latter structures are crucial for the transport of oxygen to the arterial blood and the exchange of carbon dioxide from the venous blood supply. One of the biggest challenges in solving the puzzle of the toxicity of nanotechnology will be to evaluate the toxicity of nanomaterials to the respiratory system.

Another important factor affecting toxicity is particle shape. Nanomaterials can be all types of shapes: amorphous, rods, wires, sheets, spheres, horns, dendrimers ... the list can be as long as the imagination of the inventor or engineer seeking a new product or function. It is already known, for microscale particles such as asbestos, exhaust fumes, or smoke, that shape strongly influences the toxicity due to particle-surface-catalyzed reactions or the induction of stress, such as lipid oxidation, stress proteins, or ROS.

The particulate nature of nanomaterials also limits their distribution in the food chain. Should these materials make their way into the environment in significant amounts, they may bioconcentrate to some degree; however, it is anticipated that they would not bioaccumulate or biomagnify in the food chain because they are still solid particles and may not become a truly dissolved species (which is a prerequisite for conventional toxics today, particularly in aquatic systems where macroinvertebrates and fish are exposed on a constant basis and linked via a food web). Colloids, humic and fulvic acids, and hydrophilic acids are in the same size range (as may be some naturally occurring nanomaterials, such as volcanic dusts and silts), yet they do not biomagnify. Chemicals like dioxins/furans, polychlorinated biphenyls (PCBs), methyl mercury, perfluorooctanoic acids (PFOAs, an ingredient of Teflon™), and other persistent, bioaccumulative and toxic contaminants require both a long-term residence in an aquatic system and a high order of fugacity in order to accumulate and biomagnify up a food chain. The tendency for nanomaterials to aggregate and sorb onto environmental media limits their bioaccessibility. Although it is possible, it is therefore improbable that nanomaterials would pose a risk to the environment as a result of a passive cumulative mechanism. An exception may occur if a nanomaterial contains elements or compounds that are already known to be either extremely toxic or biomagnify, such as mercury, selenium, or highly halogenated substances.

The composition of a particular nanomaterial also is very important in three respects. First, the characteristics of a nanomaterial can differ from laboratory to laboratory or from manufacturer to manufacturer. For example, it is already known that single- or double-walled carbon nanotubes (SWCNTs or DWCNTs, respectively) can differ in size, shape, and even composition, depending on the process and/or manufacturer that produced the material [5]. It therefore can be difficult to generalize bioassay results.

Second, many bulk nanomaterials contain impurities or byproducts that can significantly influence toxicity to an organism in the laboratory [5]. Similar to the production of new materials in the early to mid-20th century, the production of new nanoproducts differs from country to country, and byproducts may be introduced inadvertently that vary in content and concentration between manufacturers. Work by Plata et al. [9] illustrates this point. They evaluated the co-products of nanotube synthesis by testing various samples of commercially available, purified carbon nanotubes. Samples of SWCNTs contained iron, cobalt, and molybdenum (used to catalyze nanotube synthesis) at 1.3 to 4.1% total metals. The samples also variously contained chromium, copper, and lead at 0.02 to 0.3 parts per thousand. Such impurities could affect the toxicity of a sample of SWCNTs.

Third, some nanomaterials contain fundamentally toxic materials. A recent *in vitro* study using human lung epithelial cells [10] showed that cobalt and manganese entering the cell as nanoparticles showed eight times the toxicity of their respective water-soluble metal salts, purportedly because the latter, as ions, could not enter the cells. This so-called "Trojan-horse" mechanism also may operate with quantum dots produced for medical applications, which are essentially spherical heavy metal alloys coated with a material such as an immunoreactive protein intended to have a specific biological activity. If white blood cells engulfed these quantum dots, the coating could be broken down by degradative enzymes and the heavy metals released into the cytoplasm of the cell. The central core of the quantum dot then becomes bioavailable and therefore able to manifest toxicity to various components within the cell.

The design of experiments that measure toxicity also can influence the results. Just as with traditionally toxic materials, the form used for dosing a nanomaterial can throw into question whether an experiment is really scientifically valid. If a nanomaterial is practically insoluble in water, then many of the doses used in experiments may not be applicable to real-world situations. In fact, one can find studies reported in the literature that use doses or concentrations that may not be realistic should a nanomaterial enter a waste stream. For example, C60 fullerenes are very insoluble in water. To test the toxicity of fullerenes, researchers have used a successive series of water-insoluble solvents or other artificial means (as discussed in Section 8.4.1) to get them into aqueous suspension. Consequently, many researchers question, as they have for decades about conventional toxic compounds, "Will studies performed in the laboratory be applicable to what might happen in the field?"

Concerns about the toxicity of nanomaterials can be put in a broader perspective. With regard to aquatic systems, one group of researchers [11] stated that "[t]he increasing worldwide contamination of freshwater systems with thousands of industrial and natural chemical compounds is one of the key environmental problems facing humanity." This statement does not acknowledge the fact that natural waters have some ability to self-purify [12]. Ordinary processes that are always at work in nature naturally cleanse the water column: oxygenation of running waters, sorption of pollutants by suspended sediment and subsequent filtration by wetlands, complexation by particulate or dissolved organic matter, microbial mineralization of pollutants, and purification by filter-feeding organisms. Thus, any discussion of potential environmental effects of nanotechnology must consider the fate and transport of those materials in the environment, which may limit an organism's exposure.

8.2.2 Exposure to Nanomaterials

The sources and routes of exposure to nanomaterials are discussed below, as are the natural defenses that limit the dose to organisms once exposed. Two key factors can limit exposure. First, most nanomaterials are expensive to produce. To prevent waste and therefore loss of capital, manufacturers can carefully contain their products. Sound economics therefore can help an industry police the life cycle of its own product and thereby limit exposures. Second, many nanomaterials form much larger agglomerates, which would eventually settle out of the atmosphere or a surface waterbody onto soil or sediment. Over time, these agglomerations might bind irreversibly to these matrices.

8.2.2.1 Sources and Routes of Exposure

Various authors have developed conceptual models, some complex, of how nanomaterials might work their way into the terrestrial environment. The most obvious source, based on historical precedent, would be via emission from an industrial stack or hood ventilation system. Nanomaterials' small size precludes them from behaving like their microscale counterparts (e.g., fibers of asbestos, fiberglass, cotton). They are thus expected to behave more similarly to a gas, dissipating via advection and diffusion processes, and thus decreasing logarithmically in concentration with distance from a source. Depending on weather conditions, the nanomaterials or nanoparticles could either be carried aloft, possibly high up into the stratosphere, or, be washed down to the surrounding soils or waterbodies during a rainstorm.

For terrestrial receptors to be exposed to airborne nanomaterials, a source would have to be fairly close by for exposure to be probable and, even then, fluctuations in meteorological conditions would facilitate periods when animals whose home range fell on the upwind side of a potential air source were not exposed.

Similar to traditionally toxic materials, concentrations in soils would have to be relatively high (high part-per-million to percent range) to overcome the fate and transport processes that tend to ameliorate toxicity over time. Adsorption to and reactions within the soil matrix are anticipated to cause nanoparticles to eventually degrade, become less bioaccessible, or become less biologically active than the parent material. Because like dissolves like, carbon-based nanomaterials would, based on what we know about the behavior of other carbon-based compounds, bind to the organic fraction of the soil. The smallest nanomaterials could be bound up by irregular surface micropores of the soil matrix (unless the concentration of the nanomaterial exceeds the sorptive capacity of the soil). Future research, particularly experiments employing many different types of soil matrices, will be able to resolve whether this phenomenon will occur with carbon-, metal-, or metalloid-based nanomaterials.

Nanomaterials also can enter the environment through wastewater discharges, whether from aqueous industrial waste streams, effluent from wet scrubbers used in air pollution control, or in domestic wastewater. The latter is discussed further in Chapter 7.

8.2.2.2 Exposure and Dose

Individual organisms can be exposed to nanomaterials in their environment via ingestion, dermal contact, or inhalation. Each of these routes of exposure is discussed below, as are the natural defenses that organisms can employ to reduce the effective dose.

Oral exposure of terrestrial organisms is anticipated to be low. One reason for this is the known selectivity of the intestinal villi and, if absorbed, the effectiveness of the hepato-biliary system in eliminating particulate foreign matter from the body. Other reasons pertain specifically to terrestrial organisms. Their exposures to nanomaterials in soils are expected to be low because, unless waste disposal practices are egregious or soils are very close to a source, nanomaterials would become sorbed to micropores in the soil matrix and thereby rendered unavailable to the organism. Alternatively, they might be diluted by the soil matrix if water solubility were higher and the nanoparticles were to percolate down through the various soil horizons. With the exception of invertebrates such as earthworms that consume soil to extract nutrients, most soil-dwelling animals (e.g., shrews, mice, voles, gophers, etc.) do not, inadvertently, consume much soil (typically less than 1 or 2% of the diet; see U.S. *EPA's Wildlife Exposure Factors Handbook* [13]). Further, with a few exceptions such as metal oxides, most of the nanomaterials being produced are difficult to get into suspension and will therefore form agglomerates or precipitates, which are anticipated to become part of the soil matrix and therefore unavailable for biological uptake into an organism if the soil were inadvertently consumed.

The least probable exposure pathway will most likely be dermal, for several reasons. First, with the exception of certain invertebrates, such as earthworms, many aquatic organisms and the vast majority of terrestrial organisms have a line of defense above and beyond the dermis/epidermis layer. Fish scales overlap and, because they overlap in the same direction as the general motion or movement (forward) of the fish, the probability of dissolved nanomaterials being absorbed across the integument of the animal is anticipated to be relatively low. Different terrestrial organisms have different lines of defense. Mammals have thick coats of fur. Birds have layer upon layer of down and feathers that, microscopically, form unique interlocking networks that would act as an effective external barrier. Reptiles have thick, horny overlapping scales. Most insects (the vast majority of which are beetles) have a sclerotized dermal layer that strongly resists both physical and chemical attack. Because of their extremely small size, one might anticipate nanomaterials passing through this first line of defense. In short, nature has equipped most ecological receptors with layer upon layer of fur, feathers, scales, and/or sclerotized exteriors with coatings such as oils, fats, and waxes that will act as innate dust collectors. The effectiveness of such dust collectors depends in part on a physical phenomenon that affects the behavior of nanoparticles. Nanomaterials are subject to the random movement of adjacent molecules, a phenomenon called Brownian motion, which will increase the probability that it will encounter, and collide with, a filtering mechanism. This process is called diffusional capture [14] and appears to be effective for traditional particles less than 0.3 micrometers (μm) in size.

With the exception of aquatic or semi-aquatic organisms that may have a semipermeable dermis, such as amphibians, the respiratory system is expected to be the most vulnerable target organ. In terrestrial animals, nanomaterials may pose the greatest risk to the lung, as they can be transported like a gas to reach the alveoli in the deepest portion of the lungs. In aquatic organisms, nanoparticles may be absorbed as water is passed over gill membranes at a fairly rapid rate to extract the dissolved oxygen that is absolutely necessary to sustain the life of an individual organism.

8.2.3 SUMMARY

A host of factors will determine both the degree of exposure and the toxicity of nanomaterials to either terrestrial or aquatic receptors: the type of environmental receptor, its habitat, the duration of exposure, age, gender/sex, sensitivity or tolerance, adaptive mechanisms, and the composition, size, shape, surface area, solubility, and concentration of the nanomaterial in question. The challenge in solving the puzzle is considerable. Technology will have to rise to meet the problem of dosimetry (i.e., generating and/or measuring airborne nanomaterials). No standard metrics currently exist for quantifying the inhaled dose (particles/m^3?, surface area/m^3?, mg/m^3?). Current research programs are not universally aligned with regard to testing protocols. Finally, because of the explosion of new materials, combined with the current lack of information on how different nanomaterials behave and/or enter the body, there will be considerable uncertainty in the use of current predictive models such as physiologically based pharmacokinetic models.

The remainder of this chapter discusses the effects of nanomaterials on terrestrial and aquatic receptors. For terrestrial receptors, inhalation will be the key exposure pathway and the lung will be the key target organ, should nanomaterials enter the general environment via air. The brain also may be a target if uptake occurs through the olfactory nerves. Similarly, for aquatic receptors, water will be the obvious route of exposure and the respiratory system, namely the gills (whether they be internal gills of a fish or the external gills of some types of benthic invertebrates), are expected to be the key target organ.

8.3 ANTICIPATED HAZARDS TO TERRESTRIAL ECOSYSTEMS

At the lower levels of the food web, some nanomaterials appear to possess potent antibacterial properties [15–17], particularly materials containing silver. Researchers have exposed microorganisms (*Escherichia coli*) to nanomaterials containing silica, silica/iron oxide, and gold to examine the antibacterial response, but the growth studies have "indicated no overt signs of toxicity" [18]. Similarly, exposure of a soil microbial community to C60 fullerenes had little impact on the structure and function of the community and associated microbial processes [19]. Fullerenes in water suspensions, however, "exhibited relatively strong antibacterial activity" [20], with fractions containing smaller aggregates showing higher toxicity even though the "increase in toxicity was disproportionately higher than the associated increase in putative surface area." Aqueous suspensions of SiO_2, TiO_2, and ZnO, however, showed strong antibacterial activity (*Bacillus subtilis*), apparently through the generation of ROS [7]. The study conclusions "highlight the need for caution during use and disposal of such manufactured nanomaterials to prevent unintended environmental impacts."

Few studies to date have examined the results of dermal exposure, but one study exposing both human and rabbit skin to fullerene soot containing carbon nanotubes using patch tests [21] could not find that the mixture posed any risks. Another study used six different types of quantum dots under 12 nm in size [22] and coated with neutral, anionic, or cationic shells. The results showed penetration of intact porcine skin using *in vitro* flow-through diffusion cells at occupationally relevant doses. Full penetration could not be confirmed as the perfusate was negative for the detection of quantum dots, with a detection limit of 0.5 to 1.0 nm. The authors state that the skin was permeable to these structures in that they "penetrated the stratum corneum and localized within the epidermal and dermal layers by 8 hours," but others (carboxylic acid coated) were less effective. In any event, although intact skin may be a potential pathway for the absorption of nanomaterials, the potential for significant exposure via skin, at least for terrestrial organisms, appears to be the least probable with regard to the three available exposure pathways.

Inhalation may be the most significant exposure pathway for terrestrial organisms should an ongoing release of nanomaterials occur. The concept of ongoing release is important because these ultrafine materials, like their larger fiber counterparts, will only induce a significant pathology such as inflammation, production of biologically active substances by the RES, fibrosis, or calcification upon *chronic exposure*. Ironically, we may have already performed this type of an experiment in the real world, as urban air pollution undoubtedly contains particulate matter in the sub-micrometer range [23]. Particulate matter (in the form of "PM_{10}," or particulate matter that will pass through a 10-µm filter), principally in the form of exhaust fumes and dusts generated by the natural activity of urban life, was (and still is) a major cause of pulmonary disease in urban and suburban environs.

Most research is still in the early phases. With a few exceptions, most of the findings with regard to respiratory pathology following exposure of laboratory animals are not that different from studies performed using more traditional toxicological testing of inhaled particulate materials. Symptoms include fibrotic reactions such as granulomas, which are nonspecific lesions in response to solid matter in tissue; an increase in number and/or activity of macrophages; oxidative stress-related inflammation (usually due to the formation of short-lived but reactive molecules); tumor-related effects in rats (although this response may have been due to "overload conditions"); and a quite unique response to nanoparticles, which is their uptake by the olfactory epithelium into the brain [24]. Nanoparticulate translocation to other areas of the body appears to be specific to the unique properties of each individual nanomaterial (i.e., composition, size, shape, surface area, water solubility, and tendency to form aggregates).

8.4   ANTICIPATED HAZARDS TO AQUATIC ECOSYSTEMS

Scientists most likely will use bioassay techniques, based on years of experience with dissolved chemicals, to evaluate the aquatic hazards of nanomaterials. This section opens with a brief discussion of those techniques and their limitations with respect to nanomaterials. It then discusses the toxicity of six target nanomaterials: carbon black, fullerenes, carbon nanotubes, silver, zero-valent iron, and titanium dioxide. Throughout, this discussion refers to the summary of literature in Table 8.1.

TABLE 8.1

Effects of Nanomaterials on Different Species of Aquatic Organisms

Species	Size or Diameter	Exposure or Dose	Endpoint(s)	Effect(s)	Commentary	Ref.
Water flea (*Daphnia magna*)	30 nm filtered; 100–500 nm unfiltered	0.2, 1, 2, 5, 6, 8, and 10 mg/L TiO_2	% Survival	LOEC >0.2 but <2 mg/L (6 and 9% mortality, respectively).	Filtered: 6 mg/L TiO_2 (empirical LC50 (unfiltered (no mortality >9%).	[4]
Water flea (*Daphnia magna*)	~30	2 mg/L TiO_2	Change in hopping frequency, heart rate, and appendage movement	No effect on hopping frequency, heart rate, or feeding appendage movement.	TiO_2 filtered before *Daphnia* exposure.	[5]
Water flea (*Daphnia magna*); algae (*Desmodesmus subspicatus*)	Product 1: 25 nm Product 2: 100 nm (100% crystalline anastase)	0–50 mg/L TiO_2	% Mortality and phototoxicity	Product 1 and 2: LOEC (12.5 mg/L and "no effect" (50 mg/L), respectively, for green algae. Daphnids show no dose-response curve (phototoxic effects appear to persist after removal of light source).	Concentrations of TiO_2 are not environmentally relevant (>10 mg/L rarely seen for any chemical). At this exposure level, physical effects are suspected to cause interference with Daphnid respiration.	[7]
Carp (*Cyprinus carpio*)	21 nm	160 mg TiO_2, 97 µg/L Cd	Bioaccumulation: (2, 5, 10, 15, 20, 25 days)	TiO_2 had stronger sorptive capacity for Cd than naturally suspended sediment. TiO_2 greatly enhanced the bioaccumulation of Cd into carp (146%).	Positive correlation between TiO_2 concentration and Cd in viscera and gills of fish.	[8]

TABLE 8.1 (CONTINUED)
Effects of Nanomaterials on Different Species of Aquatic Organisms

Species	Size or Diameter	Exposure or Dose	Endpoint(s)	Effect(s)	Commentary	Ref.
Zebrafish (*Danio rerio* – embryos)	5-46 nm (average 11.6 nm)	0.04, 0.06, 0.07, 0.08, 0.19, 038, 0.57, 0.66, 0.71 nM Ag (8 cell embryos for 120 hr post-fertilization)	% Survival; abnormal development of fins, tail, spinal cord, heart, yolk sac, head, and eyes	LOEC (% survival; tail and spinal cord flexure and truncation) appears to be 0.08 nM of silver nanoparticles; developmental sensitivity: finfold > tail and spinal cord > cardiac > yolk sac edema > head edema > eye malformation.	Comparative toxicology (e.g., relative to silver ion) difficult to discern due to dosimetry and lack of statistical analysis of data (also unclear whether residual silver may be having some effect). Unclear what the "critical concentration" of 0.19 nM silver nanoparticles means.	[18]
Zebrafish (*Danio rerio* — embryos)	10–1000 nm	120, 240 mg/L CB	% Hatch; length; head/trunk angle	None	—	[1]
Protozoa (*Stylonychia mytilus*)	60–100 nm	0.1–200 mg/L DWCNT	Inhibition of growth	Dose-dependent growth inhibition over a period of 5 days with a LOEL of 1 mg/ L. Hormesis effects on viability >110% for 0.1 mg/ L and >100% for 5 mg/L. Damage to the macronucleus and external membrane of the cells. MWCNT (100 mg/L) stimulated.	Ultrastructure showed MWNTs also inside cell mitochondria. Suggested *Stylonychia* could play a "significant role" as a "bio-scavenger of carbon nanotubes" in the aqueous environment. LOEL not likely to be an environmentally relevant concentration.	[11]

Organism	Size	Dose	Endpoint	Results	Comments	Ref.
Protozoa (*Tetrahymena pyriformis*)	40–100 nm	MWCNT = 0.1, 0.2, 2.0, 10, 50 100 mg/L	Growth of *Tetrahymena pyriformis* in either peptone yeast extract or filtered pond water	Growth in peptone yeast extract medium, but inhibited the growth of the cells (50–70% of controls) when cultured in filtered pond water.	Effects seen at highest dose (100 mg/L) most likely not environmentally realistic.	[16]
Copepod (*Amphiascus tenuiremis*)	Fluorescent <18 nm; purified "AP" 50–100 nm (10-nm diameter)	0–10 mg/L SWCNT	Life-cycle mortality, development, and reproduction	No effects with purified SWCNTs; "as prepared" (AP) SWCNTs showed effects on all 3 endpoints, with an LOEL of 10 mg/L; fluorescent fractions showed life-cycle effects at all dose levels.	Copepods were exposed to "as prepared" SWCNTs, purified (electrophoresis), or fluorescent (<18 nm) nanocarbon byproducts. Conclude effects are size-dependent (smallest most toxic).	[12]
Water flea (*Daphnia magna*)	1.2 nm diameter; length not specified	2.5, 5, 10, 20 mg/L SWCNT	*Daphnia* ability to ingest and modify lipid-coated SWCNTs.	*Daphnia* were able to ingest and utilize the lipid coating of SWCNTs as a nutrient source; LOEC for mortality (20%) 10 mg/L	Purity of SWCNTs was 85%; *Daphnia* had a "strong effect on the solubility of SWCNTs" (decreased % remaining dissolved, apparently by ingestion).	[13]
Zebrafish (*Danio rerio* – embryos)	11	20–360 mg/L SWCNT	% Hatch; length; head/trunk angle	Negligible	Slight hatching delay at 72 hpf (attributed to low O_2 or Ni and Co).	[1]
Zebrafish (*Danio rerio* – embryos)	—	120, 240 mg/L DWCNT	% Hatch; length; head/trunk angle	None	—	[1]

TABLE 8.1 (CONTINUED)
Effects of Nanomaterials on Different Species of Aquatic Organisms

Species	Size or Diameter	Exposure or Dose	Endpoint(s)	Effect(s)	Commentary	Ref.
Rainbow trout (*Oncorhynchus mykiss*)	<5 nm with SDS; 50–100 nm "strands" without SDS	0, 0.1, 0.25, 0.5 mg/L SWCNT for up to 10 days	Gill ventilation rate; gill pathology; gill mucus secretion; blood chemistry; tissue metals.	No change in tissue metal concentrations or blood chemistry. Dose-dependent rise in gill ventilation rate, gill pathology, and gill mucus secretion. Increase in GSH in gills and liver.	Conclude SWCNT are respiratory toxicant to trout, but stock solutions of SWCNT dispersed by either sonication or SDS, so exposure concentrations may not be environmentally relevant.	[10]
Zebrafish (*Danio rerio* – embryos)	MWCNT not cited. Colloidal nanosilver (5–20 nm)	MWCNT (0.001–10 mg/L (series)). Colloidal Silver (1, 10, 100, 1000, 5000 g/L)	Quantitative RT-PCR to analyze expression patterns of 8 separate detoxification genes.	Nanosilver had "clear effects on the expression of most genes in a dose-dependent manner." Changes induced by MWCNTs "were much lower," suggesting a reduced toxicity at tested concentrations.	MWCNTs were "ground, suspended in an aqueous solution and ultrasonicated for 6 hours." Colloidal nanosilver purchased directly from Polytech & Net Gmbh (10,000 ppm stock solution).	[15]
Largemouth bass (*Micropterus salmoides*)	30–100 nm	0.5–1 mg/L BF	Lipid peroxidation, protein oxidation, and total GSH.	Increase in brain lipid peroxidation (but no dose/ response). Slight decrease in gill GSH.	Water clarity increased in the C60 dosed aquaria, putatively due to inhibition of bacteria.	[3]

Species	Size	Concentration/Exposure	Endpoint	Result	Comments	Ref.
Fathead minnow (*Pimephales promelas*)	1–200 nm	500 µg/L BF for 48 hr (static renewal)	Lipid peroxidation (LPO) in brain and gill; induction of CYP2 isoenzymes in liver	100% mortality in THF-nC60 exposed fish; nC60 fish showed "no obvious physical effects after 48 hr."	Study shows use of THF as a vehicle to solubilize BF into water can confound the results of toxicity assays. Water-stirred nC60 significantly increased LPO in gill but not brain, and induced CYP2 enzymes in liver.	[17]
Water flea (*Daphnia magna*)	10–20 filtered; 20–100 unfiltered	40, 180, 260, 350, 440, 510, 700 800 mg/L BF	% Survival	LOEC (filtered) 260 mg/L (14% mortality). LOEC (sonicated) = 0.5 mg/L	C60 solvent treated prior to test; LC50 (filtered) = 460 mg/L; LC50 (sonicated) = 7.9 mg/L	[4]
Water flea (*Daphnia magna*)	10–20 nm	260 mg/L BF	Change in hopping frequency, heart rate, and appendage movement	Hopping frequency increased (3x; slight increase in heart rate and appendage movement.	Solvent pretreatment (to get into solution) of C60 prior to bioassay.	[5]
Water flea (*Daphnia magna*)	1–200 nm	*Daphnia*: 5, 10, 25, 100, 500, 1000, 5000 µg/L BF for 48 hr (BF prepped w/ and w/o THF)	48 hr LC50 (static renewal)	LC50 for *Daphnia* exposed to THF-nC60 was "at least one order of magnitude less (0.8 ppm) than for water-stirred-nC60 (>35 ppm)."	Study shows use of THF as a vehicle to solubilize BF into water can confound the results of toxicity assays.	[17]
Water flea (*Daphnia magna*)	10–20 nm	260 mg/L FD	Change in hopping frequency, heart rate, and appendage movement	Hopping frequency increased (3x; slight decrease in heart rate and increase in appendage movement).	Solvent pretreatment (to get into solution) of C60 prior to bioassay.	[5]
Zebrafish (*Danio rerio* — embryos)	~100 nm	1.5 mg/L BF	% Survival	Decreased survival (<80% at 48 hr, <45% at 96 hr).	Solvent pretreatment (to get into solution) of C60 prior to bioassay.	[2]

TABLE 8.1 (CONTINUED)
Effects of Nanomaterials on Different Species of Aquatic Organisms

Species	Size or Diameter	Exposure or Dose	Endpoint(s)	Effect(s)	Commentary	Ref.
Zebrafish (*Danio rerio* – embryos)	~100 nm	1.5 mg/L BF + GSH	% Survival	Decreased survival, but not less than 80% at all time points.	Addition of glutathione (an antioxidant) improved survival.	[2]
Zebrafish (*Danio rerio* – embryos)	~100 nm	1.5 mg/L BF	% Hatch	0% hatching rate at 60 hr and 15% at 96 hr post-fert.	Solvent pretreatment (to get into solution) of C60 prior to bioassay.	[2]
Zebrafish (*Danio rerio* – embryos)	~100 nm	1.5 mg/L BF + GSH	% Hatch	35% hatching rate at 60 hr and 70% at 96 hr.	Addition of GSH (an antioxidant) improved hatchability.	[2]
Zebrafish (*Danio rerio* – embryos)	~100 nm	1.5 mg/L BF	Heartbeat; pericardial edema	Heart rate (50% of controls); pericardial edema increased sharply between 84 and 96 hr.	Solvent pretreatment (to get into solution) of C60 prior to bioassay.	[2]
Zebrafish (*Danio rerio* – embryos)	~100 nm	1.5 mg/L BF + GSH	Heartbeat; pericardial edema	No slowed heart rate <48 hr; low rate (30%) of edema w/GSH.	Addition of GSH improved heart rate and decreased pericardial edema.	[2]
Zebrafish (*Danio rerio* – embryos)	~100 nm	50 mg/L F	% Survival; % hatch	None	85% purity (<15% C70-OH).	[2]
Zebrafish (*Danio rerio* – embryos)	~100 nm	50 mg/L F	Heartbeat; pericardial edema	None	85% purity (<15% C70-OH).	[2]

Species	Size	Concentration	Endpoint	Observations	Notes	Ref
Zebrafish (*Danio rerio* – embryos)	<220 nm	25%, 20%, 10%, 5%, 1%, 0% (vol/vol) of a 25 mg/L suspension BF	% Mortality and gross observations	Above 5% THF-C60 suspended in water: lethargy; arched backs; yolk sac and pericardial edema.	Article shows that the breakdown products of the vehicle used to suspend BFs in water (tetrahydrofuran) caused fish toxicity.	[6]
Japanese Medaka (*Oryzias latipes*)	39.4–42,000 nm	1–30 mg/L LB	Bioaccumulation of lipid-soluble latex beads	No mortality at 1 or 10 mg/L; chorion and oil droplets had higher concentrations of LBs; in adult, LBs preferred gallbladder > intestine > gonads > gills > liver/kidney (spleen/lungs were negative).	Embryos accumulated LBs principally in the yolk and gallbladder.	[14]

Note: SWCNT: single-walled carbon nanotubes; DWCNT (or MWCNT): double-walled (or multi-walled) carbon nanotubes; CB: carbon black; BF: fullerene (C60); GSH: Glutathione; F: fullerol (C60-OH); THF: tetrahydrofuran; TiO_2: titanium dioxide; FD: fullerene derivative (C60HxC70Hx); Ag, nanosilver; LB: latex beads; LOEC: lowest-observed effect concentration

8.4.1 Methodologies for Evaluating Hazards and their Limitations

From a regulatory perspective, characterizing the toxicity of xenobiotics has become easier for aquatic systems than terrestrial systems, mainly due to the fact that both acute and chronic bioassays have become standardized over time. These traditional bioassays employ a wide range of species that are easily maintained and cultured, whose life cycles are well characterized, and which will reproduce in the laboratory. Both government and private laboratories have used these species for bioassays for decades. Indeed, almost all nonnarrative water quality standards for individual compounds, whether for fresh- or saltwater ecosystems, are based on a plethora of tests on a wide variety of organisms, from engineered strains of bacteria to rainbow trout. The results of these tests are then ranked and prioritized to determine the most sensitive species, and to then obtain the most sensitive endpoint for the most sensitive life stage (usually a chronic, reproductive endpoint) for that species. Classic bioassay techniques have several limitations when it comes to testing nanomaterials, including the limited solubility of many materials.

What if a nanomaterial were to find its way into a waterbody via an industrial waste stream and be able to resist the processes of agglomeration, sedimentation, adsorption, and reaction? What would be the anticipated effects on individual organisms, communities, or populations? Unlike terrestrial ecotoxicity, where few data exist, scientists have studied the effects of several classes of nanomaterials on selected aquatic organisms. Most of these studies have focused on carbon-based nanomaterials, and critical issues such as the purity of the materials and difficulty in defining the units of dose.

Table 8.1 lists published studies that have used various test species to bioassay various types of nanomaterials. This list focuses on the six materials examined in this book: carbon black, C60 fullerenes and derivatives, single- and double-walled carbon nanotubes, silver, titanium dioxide, and zero-valent iron. In perusing these articles, it becomes immediately clear that the first obstacle many researchers had to overcome was that of getting a water-insoluble nanomaterial into solution, generally using either sonication or a solvent vehicle.

For example, several researchers [25–29] used the solvent tetrahydrofuran (THF) as an initial treatment to overcome the insolubility of C60 fullerenes. A later study conducted by Henry et al. [30] showed that the resulting toxicity may not be due to the nanomaterial being tested, but rather result from the toxic effects of the decomposition products of THF, namely γ-butyrolactone and tetrahydro-2-furanol. These treatments alone lead to the observation that, if it is so difficult to get an "insoluble" material into solution to test it on aquatic organisms, then it is unlikely that fresh- or saltwater organisms in the natural environment could be exposed to the material should it be released into an aqueous waste stream. A relatively insoluble nanomaterial would sorb or bind to other insoluble material, which would eventually be removed from the water column by natural deposition onto bed sediments. As noted in Chapter 6, however, adsorption to natural organic material can keep some nanomaterials in suspension in river water.

If this were the case, it also might be evident that the nanoparticulate in question may pose a risk to sediment-dwelling organisms, such as infaunal benthic macroinvertebrates. The authors are aware of only limited ongoing studies of the fate, transport, or effects of nanomaterials in sediments, or on benthic faunal invertebrates. Due to their extremely small size, nanoparticles would fit easily into the micropores of a sediment particle or, over time, could bind irreversibly to organic carbon and not, therefore, be bioavailable to the organism in question.

Other researchers also have questioned the purity of nanomaterials in aquatic bioassays because metals and other byproducts used in their production may be responsible for toxicity, rather than the parent nanomaterial that makes up the majority of the product. One study suggested that metal byproduct impurities caused toxicity to zebrafish embryos in a bioassay of SWCNTs. Cheng et al. [31] observed that the hatching delay of the embryos (Table 8.1) "likely was induced by the Co and Ni catalysts used in the production of SWCNTs that remained as trace concentrations after purification."

An unsettling aspect of using conventional aquatic testing protocols to identify the hazard of nanomaterials is dosimetry. No standard yet exists for defining the units of dose for nanomaterials. Table 8.1 presents doses that were cited in units of milligrams per liter (mg/L). Nanomaterials, however, are particulates that can actually be counted. Without knowing the density of the material and/or the number of particles per unit weight of the nanomaterial, the reader is left without a frame of reference against which to compare the dose. Further, surface area may influence reactivity. Thus, dose units such as nanoparticles per milliliter or surface area units per milliliter may be more relevant to toxicity than units of milligrams per liter or a molar unit such as micromolar (μM). If no standard metric(s) are adopted to ensure consistency from material to material and from test to test, then comparing two scientific studies performed using the same bioassay but with nanomaterials from different suppliers will simply be "comparing apples to oranges."

Note in Table 8.1 that the concentrations employed in *in vivo* tests range from 0.1 to 360 mg/L. For conventional materials, concentrations in the part-per-million range generally are not considered to be environmentally relevant from the standpoint of a realistic concentration in surface water. For many trace metals, for example, regulators will only be very concerned with concentrations that start to exceed 10, 50, or, for some less potent metals, 100 $\mu g/L$.

These factors are important because a research study, to be scientifically valid, must also be able to be *reproduced* by another laboratory. This includes using (1) the same type of nanomaterial, (2) the same method of test solution preparation, (3) the same dosing metric, and (4) the same toxicological testing conditions. If a second laboratory cannot reproduce the results of a study under the same environmental testing conditions, then the results of the original study will be thrown into doubt.

8.4.2 DISCUSSION OF RESULTS

In reviewing Table 8.1, it first becomes clear that the unsubstituted carbon-based compounds (e.g., single- and multi-walled carbon nanotubes, C60 fullerenes, carbon black) are difficult to get into aqueous solution and thus it is difficult to dose test organisms without introducing another test variable. Measures taken to get the nanomaterials into solution also eliminate the self-purifying mechanisms of natural waters such as adsorption, complexation, precipitation, and deposition. These nano-materials may tend to agglomerate or oxidize, particularly over time, which further affects bioassay results. In the review of the references cited in Table 8.1, it also becomes clear that nanomaterials may become contaminated with byproducts that are generated in their production.

Based on traditional laboratory aquatic bioassays, the relative hazard of the nanomaterials tested thus far appears to be in the "low to moderate" range. Given that environmental conditions are certainly not uniform in the field, the bioassay results suggest that the subset of nanomaterials including carbon-based fullerenes, single- and multi-walled nanotubes, carbon black, and titanium dioxide will pose a relatively low hazard to native aquatic organisms. One exception to this may be col-loidal silver, which is both soluble in water and contains silver, a known toxicant to aquatic organisms.

8.5 RECOMMENDATIONS FOR MANAGING THE RISKS OF FUTURE NANOMATERIALS AND THEIR PRODUCTION

As the research described in this chapter shows, some nanomaterials may be toxic to ecological receptors, but the hazard is anticipated to be limited by fate and transport processes that restrict exposure. Economics also may limit the release of nanomate-rials to the environment during manufacturing, as manufacturers seek to minimize product losses during production. The use of known toxics to produce nanomaterials may present other concerns, however, as discussed further in Chapter 11.

Although many scientists are concerned about the containment and the potential risks associated with both current and future nanomaterials, it is possible that these risks can be managed using existing technology. But the oft-quoted phrase "those that ignore history are doomed to repeat it" may hold true if there is little to no communication between scientists (particularly between toxicologists and industrial hygienists with research scientists who are inventing these materials at a rapid rate), regulators, and the public. It would be prudent to treat each new nanomaterial in the same way society treats each new chemical compound, and the only way to do that is through the use of a carefully selected battery of tests and bioassays. That said, it also will be important to ensure that an industry standard is used for the mass production of the more commonly used nanomaterials so that testing regimes and results from laboratory to laboratory can be compared with confidence.

REFERENCES

1. Treverton, G.F. 2007. Risks and riddles. In "People and Culture," *Smithsonian Maga-zine*. http://www.smithsonianmagazine.com/issues/2007/june/presence-puzzle.php.
2. Knight, T., V.G.R. Chada, S.S. Wise, et al. 2007. Cell-based assay for cytotoxic and pro-inflammatory effects of gold nanoparticles. *NSTI-Nanotech 2007*, 2. http://www.nsti.org/procs/Nanotech2007v2/7.
3. International Agency for Research on Cancer (IARC). 1996. Summaries and Evalua-tion: Carbon Black 65:149. http://www.inchem.org/documents/iarc/vol65/carbon.html. (Accessed September 29, 2007)

4. Rand, G., P.G. Wells, and S. McCarty. 1995. Figure 2 of Chapter 1. In *Fundamentals of Aquatic Toxicology: Effects, Environmental Fate, and Risk Assessment*, Ed. Gary Rand. Washington, D.C.: Taylor & Francis.

5. Bell, T.E. 2007. Press Backgrounder. In *Reporting Risk Assessment of Nanotechnology*. http://www.nano.gov/html/news/reporting_risk_assessment _of_nanotechnology. pdf.

6. Long, T.C., N. Saleh, R.D. Tilton, G.V. Lowry, and B. Veronesi. 2006. Titanium dioxide (P25) produces reactive oxygen species in immortalized brain microglia (BV2): implications for nanoparticle neurotoxicity. *Environ. Sci. Technol.*, 40(14):4346–4352. (DOI: 10.1021/es060589n)

7. Adams, L.K., D.Y. Lyon, A. McIntosh, and P.J.J. Alvarez. 2006a. Comparative eco-toxicity of nanoscale TiO_2, SiO_2, and ZnO water suspensions. *Water Res.*, 40(19):3527–3532.

8. Adams, L.K., D.Y. Lyon, A. McIntosh, and P.J.J. Alvarez. 2006b. Comparative toxicity of nano-scale TiO_2, SiO_2 and ZnO water suspensions. *Water Sci. Technol.*, 54(11-12): 327–334.

9. Plata, D.L., P.M. Gschwend, and C.M. Reddy. 2006. Co-Products of Carbon Nanotube Synthesis: Emerging Contaminants Associated with the Nanomaterial Revolution. Symposium honoring Dr. Walter Giger, Organized by J.A. Field. Symposia papers presented at the *Division of Environmental Chemistry,* American Chemical Society, Boston, MA, August 19–23, 2006.

10. Limbach, L., P. Wick, P. Manser, R.N. Grass, A. Bruinink, and W.J. Stark. 2007. Exposure of Engineered Nanoparticles to Human Lung Epithelial Cells: Influence of Chemical Composition and Catalytic Activity on Oxidative Stress. (DOI: 10.1021/ s062629t)

11. Schwarzenbach, R.P., B.I. Escher, K. Fenner, et al. 2006. The challenge of micropollutants in aquatic systems. *Science,* 313(5790):1072–1077.

12. Drinan, J. E. and F. R. Spellman. 2001. In *Stream Ecology and Self-Purification,* Joanne E. Drinan and Frank R. Spellman (eds.). Lancaster, PA: Technomic Publishing Company.

13. U.S. Environmental Protection Agency. 1993. *Wildlife Exposure Factors Handbook*. Volume I and II. Office of Health and Environmental Assessment, Office of Research and Development, U.S. Environmental Protection Agency, Washington, D.C. (December).

14. Texas A&M Engineering. 2005. Interim Guideline for Working Safely with Nanotechnology. http://engineer.tamu.edu/safety/guidelines/Nanotechnology/ NANO_SafeGuideline. pdf.

15. Cho, K.H., J.E. Park, T. Osaka, and S.-G. Park. 2005. The study of antimicrobial activity and preservative effects of nanosilver ingredient. *Electrochimica Acta,* 51(5):956–960.

16. Michielsen, S., I. Stojiljkovic, and G. Churchward. 2006. Novel Nano-Coating Kills Viruses and Bacteria when Exposed to Light. Azonano News (November). http://www. azonano.com/news.asp?newsID=3278.

17. Park, D., J. Wang, and A. M. Klibanov. 2006. One-Step, Painting-Like Coating Procedures To Make Surfaces Highly and Permanently Bactericidal. *Biotechnol. Prog.,* 22(2):584-589.

18. Williams, D. N., S. H. Ehrman, and T. R. Pulliam-Holoman. 2006. Evaluation of the microbial growth response to inorganic nanoparticles. *J. Nanobiotechnol.,* 4(3)(DOI:10.1186/1477-3155-4-3) http://www.jnanobiotechnology.com/content/4/1/3.

19. Tong, Z., M. Bischoff, L. Nies, B. Applegate, and R.F. Turco. 2007. Impact of fullerene (C60) on a soil microbial community. *Envir. Sci. Technol.*, 41(8):2985–2991.

20. Lyon, D.Y., L.K. Adams, J.C. Falkner, and P.J.J. Alvarez. 2006. Antibacterial activity of fullerene water suspensions: Effects of preparation method and particle size. *Environ. Sci. Technol.*, 40(14):4360–4366.

21. Huczko, A. and H. Lange. 1999. Fullerenes: Experimental evidence for a null risk of skin irritation and allergy. *Fullerene Sci. Technol.*, 7:935–939.
22. Ryman-Rassmussen, J.P., J.E. Riviere, and N.A. Monteiro-Riviere. 2006. Penetration of intact skin by quantum dots with diverse physiochemical properties. *Toxicolog. Sci.*, 91(1):159–165.
23. U.S. Environmental Protection Agency. 2007. Basic Concepts in Environmental Sciences, Module 3: Particulate Matter. http://www.epa.gov/eogapti1/module3/category/category.htm#less0.1.
24. Borm, P.J.A., D. Robbins, S. Haubold, et al. 2006. The potential risks of nanomaterials: a review carried out for ECOTEC. *Particle and Fiber Toxicology*. (DOI: 10.1186/1743-8977-3-11) http://www.particleandfibretoxicology.com/content/3/1/11.
25. Oberdorster, E. 2004. Manufactured nanomaterials (Fullerenes, C60) induce oxidative stress in the brain of juvenile largemouth bass. *Environ. Hlth. Perspect.*, 112(10):1058–1062.
26. Lovern, S. and R. Klaper. 2006. *Daphnia magna* mortality when exposed to titanium dioxide and fullerene (C60) nanoparticles. *Environ. Toxicol. Chem.*, 25(4):1132–1137.
27. Lovern, S.B., J.R. Strickler, and R. Klaper. 2007. Behavioral and physiological changes in *Daphnia magna* when exposed to nanoparticle suspensions (titanium dioxide, nano-C60, and C60HxC70Hx). *Environ. Sci. Technol.*, 41(12):4465–4470.
28. Zhu, S., E. Oberdorster, and M.L. Haasch. 2006. Toxicity of an engineered nanoparticle (fullerene, C60) in two aquatic species, *Daphnia* and fathead minnow. *Marine Envir. Res.*, (62(1), S5–S9.
29. Zhu, X., L. Zhu, Y. Li, Z. Duan, W. Chen, and P.J.J. Alvarez. 2007. Developmental toxicity in zebrafish (*Danio rerio*) embryos after exposure to manufactured nanomaterials: Buckminsterfullerene aggregates (Nc60) and fullerol. *Environ. Toxicol. Chem.*, 26(5):976–979.
30. Henry, T.B., F. Menn, J.T. Fleming, J. Wilgus, R.N. Compton, and G.S. Sayler. 2007. Attributing effects of aqueous C60 nano-aggregates to tetrahydrofuran decomposition products in larval zebrafish by assessment of gene expression. *Environ. Health. Perspect.*, 115(7): 1059–1063. http://www.ehponline.org/members/ 2007/9757/9757.pdf.
31. Cheng, J., E. Flahaut, and S.H. Cheng. 2007. Effect of carbon nanotubes on developing zebrafish (*Danio rerio*) embryos. *Environ. Toxicol. Chem.*, 26(4):708–716.

9 Toxicology and Risk Assessment

Chris E. Mackay and Jane Hamblen
AMEC Earth & Environmental

CONTENTS

Toxicological risk assessment, a common tool in regulatory science, projects or characterizes the potential and extent for a given situation to result in a defined adverse effect. It usually involves a consideration of an exposure rate, which is then compared to a rate related to a given toxic response. Risk, then, is quantified based on the possibility or probability of the exposure rate meeting or exceeding the rate that causes toxicity.

Both exposure and response depend on an agent's chemistry relative to its environmental transport, distribution, and fate within the target organism (pharmacokinetics), and its ability to elicit an adverse response at one or more sites or receptors (activity). Any change in the chemical disposition of an agent that affects exposure, pharmacokinetics, or activity inevitably will alter the projections of potential adverse effect and thereby the risk.

A nanomaterial is a particulate manifestation of one or more identifiable chemicals combined as an insoluble entity in its medium of transport. Because covalent interactions would negate the particle's identity as a nanomaterial, interactions with the suspending medium usually involve only weak or Coulomb forces. By definition, nanomaterials range in size from 1 to 100 nanometers (nm). The uniqueness of nanomaterials is based on the fact that they present an environmentally or toxicologically reactive entity with a multi-atomic or multi-molecular surface associated with non-surface constituents. The surface properties of these particles often differ from their molecular form with regard to photo- and electrochemistry as well as reactive thermodynamics [1]. Furthermore, their size imparts to nanomaterials a potential for environmental and pharmacokinetic distributions that differ from both larger particulate and smaller molecular forms. These departures can significantly impact the risk assessment by altering or even negating inherent assumptions regarding both exposure and toxicological response.

At the time of publication of this work, the understanding of the actual exposure and toxicology of specific nanomaterials was still in its infancy. To aid in the progress of risk assessment for nanomaterials in the environment, this chapter concentrates first on aspects of the assessment process that would be specific and unique to nanomaterials, and second on how to integrate these considerations within a risk paradigm useful for the evaluation of human and ecological safety. (Note that Section 9.8 lists the symbols used in the mathematical models in these discussions.) The chapter concludes with a brief review of the current knowledge base.

9.1   RISK ASSESSMENT AND NANOMATERIALS

Risk assessment is the quantitative analysis intended to predict the magnitude of a response as the result of an event. In this case, the event is the presentation of a nanomaterial at a given rate or concentration, and the response is a physiological impairment within a defined receptor. This type of toxicological risk assessment originated in medical and clinical practices. Its use has since expanded to quantify situations involving matters ranging from product safety to environmental pollution.

Application of toxicological risk assessment to nanomaterials will not require a significant change in the standard paradigms. However, it will entail new considerations that previously were either insignificant or could be reasonably generalized using conservative or equilibrium-based assumptions. For nanomaterials, such generalizations could be extremely imprecise. Hence, considerations such as partition-independent penetration, inflammatory and sensitivity reactions, and disequilibrium dynamics will be required to accurately quantify risk.

9.1.1   EFFECTS OF STERIC HINDRANCE

Nanomaterials, like ultrafine particles, do not necessarily follow the same toxicological paradigms as molecular toxicants. Differing routes and altered potential for absorption can result in different exposures. The toxicological response to particulate toxicants may not always follow the concentration gradient because of steric limitations resulting from the particle size. Steric limitations arise when a physiological barrier retards or prohibits the movement of the material, regardless of the concentration gradient. Therefore, a nanoparticulate form of a material may have no effect, whereas a molecular form may invoke toxicity simply because the larger nanoparticulate form cannot reach the site of action. Conversely, steric inhibition to transport may cause a nanomaterial to accumulate in a particular physiological region, resulting in a unique toxicological response. For example, a molecular toxicant that causes systemic toxicity may, when in nanoparticulate form, cause only toxicity at the point of environmental contact because of steric inhibition to absorption of the nanoparticle. However, risk assessment must consider variations in response. Many of the physiological barriers to particulate exposure, absorption, and even response, tend to vary greatly within the general population. This may result from physiological conditions (age, disease state, etc.), co-exposure to other environmental factors, and/or genetic predispositions. As a result, it will be important to quantitatively consider this variability when selecting toxic endpoints and predicting the proportional response of the exposed population in any risk assessment.

9.1.2 INFLAMMATORY AND IMMUNE-BASED MECHANISMS

The general understanding of the toxicity of nanomaterials is still evolving with, in some cases, surprising results. Initial research shows that inflammatory and immune-based mechanisms of toxicity may be particularly important for nanomaterials. For example, the most significant toxicity currently attributable to a nanomaterial results from exposure to single-walled carbon nanotubes. Such exposure can cause pulmonary inflammation manifesting in granuloma and fibrosis. The relative importance of inflammatory and immunogenic responses can significantly complicate risk assessment because such responses, as an adverse effect, vary widely within the general population. The same toxicant exposure could elicit responses in different people ranging from no effect to life threatening.

Intrapopulation variability confounds attempts to quantify the probability and magnitude of immunogenic or inflammatory response. Sensitivity may not only vary with genotype, but also with factors such as age and exposure history. Thus it is very difficult to predict. The *a priori* identification of sensitive sub-populations will be challenging and may require the development of screening methods not currently employed in environmental risk assessment. The significance of this variability will depend on the relative prevalence of a predisposition to response within the general population. Current advances in toxicogenomics will provide the basis for characterizing sub-population sensitivities and is likely to become a significant consideration in the risk assessment of nanomaterial exposure.

9.1.3 CRITICAL VARIABLES

The toxicity of a nanomaterial, as with any agent, depends on the chemical properties that determine its potential interactions with various cellular targets in an organism. Defining exposure as the presentation of the potential toxicant to the target organism at the environmental boundary (*ex integument*), the toxicity then can be considered as the intersecting functions of absorption, distribution, response (which is the combination of damage and repair relative to homeostasis), metabolism, and elimination. The manifestation of a toxic response often varies with the route of exposure, depending more on the amount, barriers to absorbance, and transport of the toxicant than on the actual activity of the toxicant itself. Examining toxicity based on routes of exposure isolates the differential responses and segregates subpopulations with respect to activities incurring exposure and in terms of an easily measurable dose factor. The principal routes of exposure considered here are oral ingestion, dermal absorption, and inhalation.

9.2   EXPOSURE AND EFFECTS THROUGH INGESTION

Ingestion and inhalation, rather than absorption through the skin, are the most likely method of direct exposure to nanoparticles. (See Section 9.4 on inhalation exposure.) There are two important considerations in assessing the risk related to the ingestion of nanomaterials. The first is the potential direct toxicity resulting from contact with the digestive epithelium. The second is the potential for the nanomaterial to enter the blood circulation (central compartment) via the digestive tract and thereby be systemically distributed.

Increasing the size of a compound or particle decreases its ability to cross a cellular barrier. This can result from steric hindrance (the particle is too large to physically fit through a pore or space) or thermodynamics (the rate of movement is too slow to be of consequence).

The epithelium of the digestive tract contains tight junctions that limit the size of materials that can pass between cells to enter the central compartment. Particles with an effective diameter greater than 4 nm cannot pass between the cells [2] and therefore must undergo cellular transport, either passively or actively. Active transport, via channel transport or endocytosis, is subject to the limited capacity of the cell to transport material. Passive transport is driven by the diffusion gradient and is subject to the permeability of intervening membranes. Passive cellular transport can be considered a two-step chemical reaction. First, a particle dissolved in digestive fluids partitions and dissolves in the cell's lipid bilayer membrane. Second, the particle partitions and dissolves in the cytosolic medium. This process also is subject to thermodynamic limitations. To predict the rate of absorption for a nanomaterial with a variable size and surface behavior requires that this two-step reaction be broken into its components.

9.2.1   DIFFUSION

The introduction of a molecule into the lipid bilayer is an endothermic process. The energy necessary to initiate the process is provided by the combined partition gradient (i.e., differential affinity of a solute for an aqueous vs. non-aqueous medium) and concentration gradient, and is released once the compound leaves the membrane. The larger the compound, the more energy is necessary for it to transfer from the aqueous phase into the lipid phase of the bilayer. This may be considered in terms of the probability of a hole forming in the bilayer large enough to accommodate the compound: the larger the compounds, the lower the probability an appropriate sized hole will be formed to accommodate the nanomaterial, and the slower its passage into the membrane.

Lieb and Stein [3] described a model for determining the diffusion rate of materials through a bilayer based on size-dependent steric hindrance. Briefly, the permeability of the bilayer to a given compound (P) is the product of the partition coefficient of a solute relative to the aqueous medium (k_{mem}) and the diffusion coefficient of the membrane (D_{mem}) relative to the diffusion distance or membrane thickness (d_{mem}) as follows:

$$P = \frac{k_{mem} \cdot D_{mem}}{d_{mem}} \tag{9.1}$$

Hence:

$$D_{mem} = \frac{P}{k_{mem}} \cdot d_{mem} \tag{9.2}$$

where d_{mem} is constant regardless of solute. Therefore, the effect of molecular size can be isolated from molecular volume (V) as the empirical relation of D_{mem} vs. V (Figure 9.1 [4]) with the following relation:

$$D_{mem} = D_{mem}^{V=0} \cdot 10^{-(m_v V)} \tag{9.3}$$

Combining the two equations above, the slope of this relation (m_v) can then be applied to determine the theoretical permeability (P) assuming a molecular volume of zero ($P^{V=0}$).

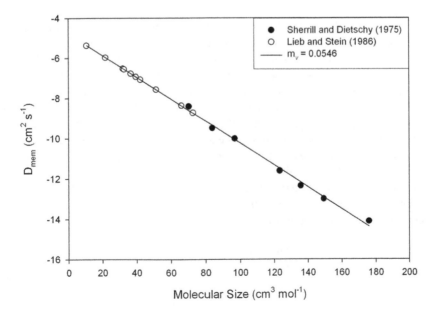

FIGURE 9.1   Size correction relation (m_v) applied to determine molecular permeability (P) from the theoretical zero-volume permeability ($P^{v=0}$).

$$P^{V=0} = \frac{k_{mem} D_{mem}^{V=0}}{d_{mem}} = P \cdot 10^{(m_v V)} \tag{9.4}$$

Lieb and Stein [3] showed that log $P^{V=0}$ correlates with log k_{ow} with a slope of 0.0546. This allows for the description of the overall permeability in terms of volume and partition as follows:

$$P^{V=0} = P \cdot 10^{(m_v V)}$$

$$P = \frac{P^{V=0}}{10^{(m_v V)}} \tag{9.5}$$

$$= \frac{10^{0.0546 \log k_{ow}}}{10^{(m_v V)}}$$

Thus, the initial influx rate (J_{mem}) can be determined as follows:

$$J_{mem} = -D_{mem} \frac{dC}{dx}$$

$$D_{mem} = P \cdot dx$$

$$\frac{dn}{dt} = -D_{mem} A_{mem} \frac{dC}{dx} \tag{9.6}$$

$$\frac{dn}{dt} = -P A_{mem} \cdot dC$$

where n is the number of particles, A_{mem} is the membrane surface area available for absorption, and dC/dx is the concentration gradient.

The diffusion model, as parameterized, predicts the trans-membrane flux from extracellular to intracellular spaces within the digestive epithelium. This, however, is expected to be initially faster than diffusion from the intercellular to the central compartment because: (1) while the permeability P is not likely to differ significantly across the epithelial cells, the microvilli on the exterior of the digestive epithelium dramatically increase the cellular surface area (A_{mem}); and (2) the initial concentration gradient from the digestive tract to the intracellular compartment is greater than the gradient from the epithelium to the central compartment.

To predict transport kinetics from the digestive tract to the central compartment, the membrane diffusion model must be coupled into a three-compartment model (Figure 9.2) to isolate the rate-limiting step as follows:

$$\frac{dn_1}{dt} = -P A_{GI} \left([C]_{IC} - [C]_{GI} \right)$$

$$\frac{dn_2}{dt} = -P A_{CC} \cdot \left([C]_{CC} - [C]_{CI} \right) \tag{9.7}$$

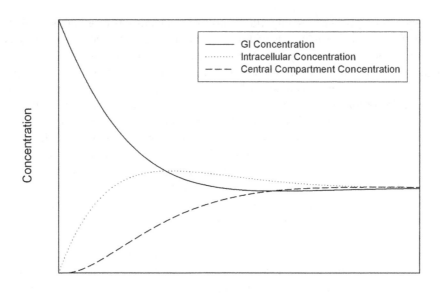

FIGURE 9.2   Time course of diffusion equilibrium across the intestinal epithelium.

where:

dn_1/dt = Rate of solute flux from gastrointestinal (GI) tract to GI epithelial cell

dn_2/dt = Rate of solute flux from GI epithelial cell to the central compartment

A_{GI} = Cellular surface presented to the GI tract

A_{CC} = Cellular surface presented to the central compartment

$[C]_{GI}$ = Solute concentration within the GI tract

$[C]_{IC}$ = Solute concentration in the GI epithelium cell

$[C]_{CC}$ = Solute concentration within the central compartment

While data are available to determine the relations of D_{mem} vs. V and $P^{v=0}$ vs. k_{ow}, one problem with this approach in relation to nanomaterials is the lack of comparable data related to the permeability to materials in an appropriate size range. While first principal thermodynamics suggests that if the original relations are accurate, the relation between P and V should hold through the nanoparticle range; the relation between $P^{V=0}$ and k_{ow} is in fact a structure/activity relationship and may not be valid in extrapolation to such large particle sizes. This data gap must be filled to understand the potential absorption and hence toxicity of ingested nanomaterials.

9.2.2   ENDOCYTOSIS

Endocytosis refers to the process of cellular transport without requiring transmembrane diffusion. It usually involves the activation of a membrane receptor that results in the invagination and separation of a membrane vessel within which the activating material is contained. The cell, in effect, engulfs the particle. For nanomaterials, endocytosis may be the most important transport mechanism because of the predicted low diffusion rates for materials with volume on the order of hundreds to thousands of cubic nanometers. Endocytosis tends to follow the concentration gradient, in that high exogenous particle concentrations result in high rates of endocytotic transport. However, the capability to initiate endocytosis is chemical and cell-specific, and the kinetics do not follow a diffusion relation. This necessitates the use of specific empirical expressions for the derivation of P that cannot be derived thermodynamically.

Nanoparticles have been shown to be transported by endocytosis into the central compartment with a size cut-off of about 300 nm [5]. It is known that particulate matter is transported from the intestinal lumen into the lymphatic system via Peyer's patches that contain specialized endocytes called M-cells. Uptake via the intestinal epithelium or intestinal lymphatic tissue results from an induced cellular response and therefore would be expected to vary by nanomaterial size, partition characteristics, and charge distribution.

Few data describe the potential for ultrafine or nanomaterials to impact the gastrointestinal tract. Particulate metals in high concentrations can disrupt the fluid balance in the colon. Some evidence indicates that ultrafine particles may be involved in inflammatory conditions such as irritable bowel syndrome and Crohn's disease [6]. However, a genetic predisposition appears to be required for the condition to manifest itself, thereby making population-based generalizations difficult in risk assessment. Nanoparticles of zinc have reportedly induced both contact and systemic toxicity upon ingestion [7]. However, it is unclear whether these are particle effects or the result of zinc dissolution from the particle surface.

9.3   EXPOSURE AND EFFECTS THROUGH DERMAL ABSORPTION

To date, no specific reports have indicated dermal toxicity resulting from exposure to an identified nanoproduct. However, ultrafine metal particles have been known to cause contact dermatitis, as have polyaromatic hydrocarbon-contaminated soots [8, 9].

Reportedly, nanoparticles of titanium oxide [10], transition metals [11], liposomes [12], and functionalized fullerenes [13] can penetrate through the outer layers of the skin (stratum corneum) into the viable epidermis and dermis. The rates and amounts vary with the material as well as the health of the receptor. Conditions such as age, site of exposure, and certain chronic disease conditions mediate the rate and extent of penetration. Secondary exposure factors such as vehicle, pH, and even humidity can dramatically affect particulate penetration [14]. Past research on particle penetration has involved the movement of particles through the stratum corneum via impromptu channels formed between the subsequent layers [15]. The thickness and permeability of stratum corneum varies with location on an individual. Hair follicles also may act as a conduit for the movement of materials from the environment into the dermal layers. Similar to the stratum corneum, hair follicles are also protected by a horny layer, although it tends to be thinner than that present on surface skin [14]. Studies with micro-scale titanium dioxide (TiO_2) particles indicate penetration of the epidermal layers with the greatest concentrations clustered about the hair follicles [10].

In risk assessment, dermal penetration follows the concentration gradient. However, the penetration of the stratum corneum is extremely rate limiting. As a result, an attenuating gradient forms across this layer. Studies with polysaccharide microparticles demonstrated this gradient with almost no subdermal penetration [16]. The gradient is difficult to model based on the multifactorial nature of the diffusion dynamics. Furthermore, particulate matter that does reach the epidermal and dermal layers is subject to phagocytosis by Langerhans cells and other macrophages, which results in transport to the lymphatic system rather than the central compartment. While limiting systemic exposure, lymphatic transport may result in inflammation and hypersensitization reactions not immediately associated with the point of contact with the causative nanomaterial [17].

9.4   EXPOSURE AND EFFECTS THROUGH INHALATION

Generally, most of the work regarding exposure to nanomaterials derives from concerns related to the inhalation of ultrafine particles found in certain occupational settings, as well as ultrafine aerosols resulting from combustion. Scientists have specifically linked serious chronic diseases to the inhalation of ultrafine particles. These diseases include Clara cell carcinomas (polycyclic aromatic hydrocarbons), mesothelioma (asbestos), and berylliosis (beryllium). General syndromes associated with exposures to aerosols include black lung (coal), emphysema (combustion products), and metal fume fever (zinc, tin, and other transition metals).

Relatively stable aerosols consist of a suspension of nonvolatile particles ranging from 10 nm to 25 micrometers (μm). Typically, aerosol particles less than 500 nm deposit with a pattern more like that of a gas than a particulate suspension. Hence, diffusion governs deposition and can be expected to occur throughout the respiratory tract, including the alveoli. Deposition depends on the adherence and residence time of the nanoparticles. Particles between 500 nm and 25 μm demonstrate a slow depositional pattern where the majority may be deposited in the upper airway, but some penetrate to the deep lung. Particles larger than 25 μm tend to be deposited through gravitational deposition and will settle in the nasopharyngeal region where the flow velocity is reduced [18].

9.4.1 MECHANISMS FOR ADSORPTION AND REMOVAL

The flux rate (J) from the inhaled atmosphere to the respiratory epithelium can be predicted through a modification of Fick's law of diffusion, which is expressed as follows:

$$J = -D\frac{dc}{dx} \tag{9.8}$$

where dc is the concentration gradient, dx is the distance across the concentration gradient, and D is the diffusion coefficient. In the case of inhalation, the separation distance is a function of the size and shape of the air space. Because an inhaled nanomaterial is distributed within the three-dimensional air space, concentration requires integration over the lateral and longitudinal directions based on the concentration gradient relative to a given location along the airway. This usually can be simplified by assuming the airway is composed of a series of relatively uniform passages (nasal, pharyngeal, tracheal, bronchi, bronchioles, and alveoli). With an intrinsically constant surface area (A) and radius (r_a) within each grouping, flux dynamics (dn/dt, where n is the number of particles) can be expressed based on the area of a given passage as follows:

$$\frac{dn}{dt} = DA \cdot 4\pi \int_{x=0}^{r} \frac{dc}{dx} \tag{9.9}$$

Substituting the Stokes-Einstein equation, the relation can be expressed as a solvable expression as follows:

$$\frac{dn}{dt} = \frac{2kT}{3\eta r_p} \cdot A \cdot \int_{x=0}^{r} \frac{dc}{dx} \tag{9.10}$$

where k is the Boltzmann constant, T is the absolute temperature, η is the viscosity of the aerosol, and r_p is the radius of the nanoparticle.

The diffusion of a nanomaterial from gaseous suspension to the epithelium involves not only a change in location, but also a change in state from aerosol to hydrosol within the mucous layer of the pulmonary airways. Usually, the concentration gradient, dc/dx, needs to be modified to account for the differential fugacity between the two states. However, nanoparticles have a low escaping tendency because of their high relative masses. Because nanomaterials contacting the mucosal layer will not significantly return to the gaseous aerosol, diffusion transport is, in effect, one way, such that the integral of $dc/dx = 1$. Furthermore, because of the rate of ventilation and turbulence, the cross-sectional gradient within the airway can, for the most part, be ignored. With these two assumptions, the concentration gradient can be simplified to the differential concentration between that suspended in the air stream and that suspended in the mucosal layer.

The linear nature of the airway means that at any point (y), the concentration is equivalent to the initial concentration ($[C_0]$), minus the integral of the material lost in the previous airway as follows:

$$\frac{dn}{dt} = \frac{2kT}{3\eta r_p} \cdot A_Y \cdot \left(C_0 - \int_{y=0}^{Y} \frac{dn}{dy} \right) \tag{9.11}$$

Note that the integral is based on the linear transport of air and will differ based on whether the ventilation is in inhalation or exhalation. Furthermore, the air flow velocity (v-) places a constraint on dy, and by implication A_Y, by the amount of surface area exposed per unit time as follows:

$$A_Y = CS_Y v dt$$

Hence:

(9.12)

$$\frac{dn}{dt} = \frac{2kT}{3\eta r_p} \cdot CS_Y v dt \cdot \left(C_0 - \int\limits_{y=0}^{Y} \frac{dn}{dy} \right)$$

where CS_Y is the cross-sectional area of the airway at point y, and dy is the infinitesimal of the change in position within the airway. Note that the area is expressed as a cross-section rather than as a function of radius. This is because the presence of processes (i.e., projecting portions of bone or tissue), particularly in the nasopharyngeal region, can greatly increase the potential surface area of exposure per unit time. However, in the pulmonary region of the lungs, where the airways are relatively smooth, the exposed area per unit time can be expressed in terms of $\pi r^2_a dy$. Figure 9.3 shows examples of projected deposition rates based on mass and fiber numbers for the bronchioles. As shown, the deposition rate increases with concentration and decreases with particle size (diameter).

Direct solution of this relation is difficult because of the heterogeneity of the mammalian airway. Predictions of absorption rates usually involve the construction of a three-dimensional passage model that segments differential regions of the airway based on similar diffusion properties. These models generally indicate that the number of particles deposited is inversely proportional to the size of the particle [19]. Therefore, the smaller the particle, the larger the amount absorbed as the result of higher rates of diffusion. Although counter-intuitive, the relation also suggests that the faster the air velocity, the higher the rate of absorption. But note that this results from the increase in surface area exposure per unit time, which decreases the longitudinal gradient, thereby allowing higher concentrations in deeper regions of the airway.

Upon adsorption to the lining of the airways, particulate matter is suspended in a complex mixture called the tracheobronchial mucus. Produced by both submucosal and epithelial secretory cells, the mucus comprises a mixture of glycoproteins and electrolytes within an aqueous matrix. The viscosity of the mucus varies throughout the respiratory system, thereby altering the diffusivity of nanoparticles. The mucous layer in humans continues from the larynx to the end of the first-generation bronchioles. Within the alveoli, Type II cells also produce a proteinaceous secretion similar to mucus, but usually of a lower viscosity and higher water content.

The pulmonary mucosa is part of a clearance system referred to as the respiratory conveyer. This system of ciliated cells, which lines the bronchioles and trachea, traps inhaled particulates in mucus and sweeps the laden mucosal material up and out of the respiratory tract. Rates of movement vary from about 0.6 mm/min in the bronchioles to about 10 mm/min in the trachea region [20]. The respiratory conveyor deposits most of the material in the esophagus, which may represent a significant exposure route for the ingestion of nanomaterials.

Materials with a sufficient concentration gradient to reach the alveoli are not directly subject to the mucosal conveyer because there are no cilia in the alveoli.

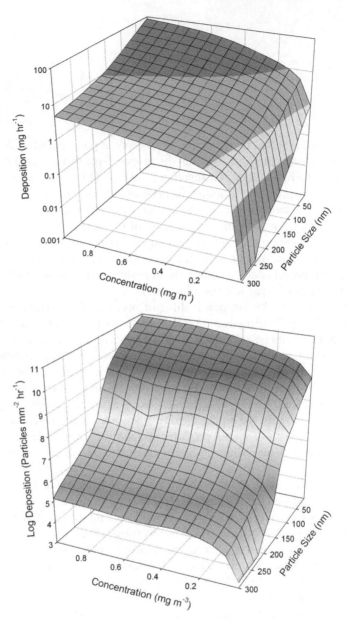

FIGURE 9.3 Depositional kinetics of nanomaterials within the human bronchioles standardized based on (a) concentration and (b) particulate number.

Three principal methods can clear nanomaterials from the alveoli. The first is diffusion based and involves the movement of nanomaterials through the Type I cells into the vascular capillary bed and the general circulation, where they are then removed by blood filtration. The second and third methods involve initial phagocytosis (engulfment) by resident macrophages. Macrophages can engulf insoluble particles from molecular dimensions up to about 1 μm in diameter [21]. The laden macrophages then can migrate vertically to the bronchioles where they are entrained in the mucosal conveyer and rapidly eliminated. Alternately, nanoparticles may be subject to endocytosis by macrophages that migrate into the lymphatic system where they are cleared via the tracheobronchial lymph nodes or the blood. This relatively slow process sometimes takes months to remove particulate material from an exposed organism.

The most important considerations in assessing the risk from exposure to nano-materials in aerosols are the size of the particles and the rates of exposure relative to the rates of response. From the discussion above, it is apparent that dispersed nano-materials will deposit all along the airways, including the alveoli. However, nano-materials, particularly the current carbonaceous materials, are rarely encountered in either the occupational or general environment as stable dispersals (see Chapter 6). The critical rate of exposure relates to the rate and magnitude of injury relative to the rates of elimination and repair. If injury resulting from exposure exceeds the airway's repair capacity as the result of inefficient removal capacity, then it can be expected that an adverse effect will ensue.

Inflammation is the most common response to fibrous or particulate material. It results from the activation of inherent defense mechanisms mediated by mac-rophages that, if over-stimulated, will result in localized cellular necrosis and loss of lung functions. A case study of the potential risk associated with single-walled carbon nanotubes follows.

9.4.2 Case Study: Inhalation of Carbon Nanotubes

Single-walled carbon nanotubes (SWCNTs) consist of a sheet of aryl carbon rings curved around on themselves so as to form a tube one layer thick. SWCNTs are typi-cally 1 to 4 nm in diameter and vary from as short as 50 nm to lengths in excess of 2 µm. Carbon nanotubes possess extremely low charge affinity compared to that of fluid media such as air and water. As such, they tend to rapidly form clumps by bind-ing to one another, particularly along their long axes. This manifests a tertiary struc-ture consisting of numerous SWCNTs in forms referred to as nanoropes. Nanoropes will associate further into groups of nanoropes referred to as tangles and will con-tinue to associate until the units become so large as to fall out of fluid suspension.

9.4.2.1 Pulmonary Toxicology

As of the date of publication, no human studies were available that evaluated the pulmonary toxicity of SWCNTs. Furthermore, animal tests for direct inhalation were not available due to the practical difficulties in isolating and collecting enough SWCNT particles to conduct these studies [22]. As such, almost all the current stud-ies are based on either *in vitro* designs using tissue explants of cultured cell lines, or exposures of whole animals using intratracheal instillation. The term "intratracheal instillation" describes a technique where researchers inject a bolus dose of a SWCNT suspension into the trachea of the test animal to distribute SWCNTs throughout the pulmonary airway by aspiration. While the intratracheal instillation method has technical limitations, it is an accepted screening test for pulmonary toxicity [22–24].

Three intratracheal instillation studies have examined the pulmonary toxicity of SWCNTs [23–25].

Lam et al. [23] instilled mice with a single treatment of 0, 0.1, or 0.5 mg/mouse SWCNT suspension in a 50 μL buffer (equal to approximately 3.94×10^6 and 1.96×10^7 fiber units per mouse, respectively). Four animals per dose group were euthanized 7 days after the single treatment; five animals per dose group were euthanized 90 days post treatment. Lam et al. reported dose-dependent lesions, primarily interstitial granulomas, in both the 7- and 90-day groups. The lesions were more prominent in the 90-day animals. Mice also were treated with quartz and carbon black (whose size range included nanoparticles). Minimal inflammation was observed in mice treated with carbon black, and moderate inflammation was observed in mice treated with quartz. Lam et al. reported that the quartz-induced toxicity was less severe than lesions induced by SWCNTs.

In a second study, Warheit et al. [25] instilled rats with SWCNTs at 0, 1 or 5 mg/kg (approximately 9.79×10^6 and 4.90×10^7 fiber units per rat, respectively). The research team euthanized and examined animals 1, 7, 30, or 90 days after a single treatment. Granulomas were present after 1 month but the lesions were neither dose dependent nor time dependent. Toxicity was not reported in rats that were treated with graphite. Based on the results, Warheit et al. Concluded that "granulomatous reaction was a nonspecific response to instilled aggregates of SWCNTs and the results may not have physiological relevance, and may be related to the instillation of a bolus of agglomerated nanotubes." Lam et al. [22] postulated that this lack of dose and time dependence reported by Warheit et al. [25] might be due to a significant portion of the instilled bolus dose not reaching the alveolar region.

Shévedova et al. [24] conducted a third study in an attempt to resolve the differences. In this study, mice were instilled with a SWCNT suspension that had been highly purified to remove metals. Mice were administered SWCNT, carbon black, or quartz at 0, 10, 20, or 40 μg per mouse (approximately 3.92×10^5, 7.84×10^5, and 1.57×10^6 fiber units per mouse, respectively). Animals were euthanized at 1, 3, 7, 28, or 60 days following a single treatment. Acute pulmonary inflammation, granulomas, and fibrosis were reported. The pulmonary toxicity was both dose and time dependent. Similar to the studies conducted by Lam et al. [23] and Warheit et al. [25], granulomas were observed at the site of deposition of SWCNT aggregates, but unique to this study was the dose- and time-dependent interstitial fibrosis in pulmonary regions away from the sites of deposition. These data indicate fibrosis induced by dispersed SWCNTs. Neither carbon black nor quartz produced granulomas or fibrosis.

Tian et al. [26] reported that SWNCTs induced the strongest adverse effect out of five nano-sized carbon materials tested on cultured human fibroblast survival. The order of toxicity from least to most toxic was as follows: carbon graphite < multi-wall carbon nanotubes (MWCNTs) < carbon black < activated carbon < SWCNT. Dispersed SWCNTs were more toxic than unrefined SWCNTs, which tended to group together in tangles, creating larger and less harmful fibrous units.

The results of these animal studies indicate that SWCNTs can induce inflammatory pulmonary toxicity in the form of granulomas that can result in fibrosis if they reach the deep lung tissue. Toxicity in the upper airway is mitigated by short residence times resulting from their rapid removal.

9.4.2.2 Risk Assessment

Assessing the risk associated with SWCNTs requires two separate considerations: (1) the ability of a material to reach a sensitive site of action, and (2) the type and magnitude of the resultant response at the sensitive site. Current studies based on intratracheal installation indicate that the sensitive site of action for SWCNTs is the deep lung tissue — specifically the respiratory bronchioles and alveoli. Indigenous macrophages engulf SWCNTs that reach this part of the pulmonary airway. This phagocytosis apparently results in an inflammation cascade similar to that seen in silicosis, which appears to manifest as a long-term or chronic condition because the macrophages bearing SWCNTs do not migrate into the upper airways as is seen with materials such as particulate graphite [25]. Chronic inflammation in the lower airway will result in damage to the underlying epithelium and the generation of scar tissue often referred to as fibrosis. Widespread damage throughout the lower airways will reduce gas transfer significantly and a condition akin to emphysema can develop. Furthermore, chronic inflammations of this type have been associated with the promotion of hyperplasias that have the potential to become cancerous [27]. However, it must be cautioned that this is not necessarily the case, and there is currently no evidence that exposure to SWCNTs will result in either cancer initiation or promotion.

Exposure of the upper airways to SWCNTs is less toxicologically significant for two reasons. First, the residence time of the SWCNTs is much shorter because particles that impact within the nasopharyngeal, tracheal, or bronchial regions of the airway are rapidly removed via the pulmonary mucous conveyer. Therefore, inflammation appears to be transient (<2 hr). Second, because the upper airway is not the site of significant gas transfer, it comprises a thicker and more robust epithelium with greater regenerative capacity and therefore is less likely to manifest significant fibrosis [28].

Consequently, the greatest potential hazard to individuals working with SWCNTs apparently would stem from exposure to materials capable of depositing within the deep lung tissue. Materials depositing within the upper airway may be acutely toxic at high concentrations but will not likely represent a serious health issue at or below exposure concentration limits established to protect the deep lung.

Initial indications from histological studies indicate that inflammation does not depend directly on the size of the SWCNT fiber impacting the pulmonary tissue [24, 25, 29]. Rather, it is the number and distribution of the impacts that results in the overall toxic response. As such, the classic risk approach of quantifying toxicity using the mass dose per unit time or unit body mass may not be appropriate. Rather, to capture the dose response, one must quantify the exposure in terms of number of fiber units per unit time, where a fiber unit is defined as any independent SWCNT, SWCNT rope, or SWCNT tangle.

Of the current animal studies described above, the study performed by Shevedova et al. [24] provides the best toxicological characterization and quantification to derive exposure guidelines. Using the endpoint of average alveolar thickness as a measure of induced fibrosis, Shevedova et al. found that a single exposure concentration of 3.92×10^5 fibers per mouse had no effect at either 28 or 60 days post exposure.

To convert this to a human exposure, the concentration in the mouse must be scaled to a human. Given that the mouse mass in the study by Shevedova et al. [24] was reported to be 20.3 g, it is possible to estimate the total lung volume (V_{tot}) as the sum of the tidal volume (i.e., the amount of air passing in and out of the lung during normal resting breath; V_T) and the anatomical dead space* (V_D) for the mouse and a 75-kg human using the algometric scaling equations of Linstedt and Schaffer [30] as follows:

$$Mouse\ (W = 0.0203\,kg):$$

$$V_T = 6.60 \cdot W_{kg}^{1.01} = 0.129\,mL$$

$$V_D = 2.20 \cdot W_{kg}^{1.01} = 0.0430\,mL$$

$$V_{tot} = 0.172\,mL$$

9.13)

$$Human\ (W = 75.0\,kg):$$

$$V_T = 6.60 \cdot W_{kg}^{1.01} = 517\,mL$$

$$V_D = 2.20 \cdot W_{kg}^{1.01} = 172\,mL$$

$$V_{tot} = 689\,mL$$

Absolute pulmonary surface area (*SA*) is difficult to determine because of the irregular geometry. However, by assuming proportional scaling to the total pulmonary volume between the mouse and human, the relative surface area for the human and the mouse can be scaled as follows:

$$\left(\frac{V_{tot-human}}{V_{tot-mouse}}\right)^{2/3} = \frac{SA_{human}}{SA_{mouse}} = 252 \qquad (9.14)$$

With an area scaling factor of 252 human to mouse, a safe dose of 3.92×10^5 fiber units per mouse can be extrapolated to 9.88×10^7 fibers per person. This level can be considered a not-to-exceed body burden for fibers less than 5 μm in effective diameter, which is the typical upper size limit for materials that are capable of reaching the deep lung.

Muller et al. [31] reported the clearance from the deep lung for MWCNTs as a constant for elimination (k_β) of 0.01 days or a half-life of 69.3 days. This assumed an inherent interaction between the MWCNT and the pulmonary physiology, and is

---

* Anatomical dead space (VD): the volume of the conducting airways from the external environment (at the nose and mouth) down to the level at which inspired gas exchanges oxygen and carbon dioxide with pulmonary capillary blood; formerly presumed to extend down to the beginning of alveolar epithelium in the respiratory bronchioles, but more recent evidence indicates that effective gas exchange extends some distance up the thicker-walled conducting airways because of rapid longitudinal mixing.

therefore not a scalable value. Using this k_β, an allowable daily exposure rate $[C]_\beta$ can be determined as follows:

$$[C]_\beta = [C]_o \cdot \left(1 - e^{-k_\beta t}\right)$$

$$(9.15)$$

where t = 1 day, $[C]_\beta$ = 9.83 × 10^5 fibers per day

Based on study results reported by the U.S. EPA [32], the ventilation rate for an adult undertaking medium activity is 1.02 m³/hr. This equates to an exposure volume of 8.16 m³ per 8-hr day. Therefore, to ensure that the total lung burden does not exceed 9.88 × 10^7 fibers, the 8-hr time-weighted concentration cannot exceed 1.20 × 10^5 fibers/m³, and the maximum 1-hr exposure should not exceed 9.64 × 10^5 fibers/m³ for fibers with an effective diameter less than 5 µm.

Currently, no published physiological or epidemiological studies describe the effect of SWCNT inhalation in humans. The available studies were performed in rodents. It is assumed in this analysis that a safe level in rodents equates to a safe level in humans. Other studies with inflammatory fibrous material appear to indicate that the cross-species comparisons are valid [33]. However, it remains an uncertainty if this relation will hold true for SWCNTs.

Histological examination of the lesions associated with SWCNTs in the deep lung suggests that the degree of granuloma formation and resulting inflammation is independent of the amount of SWCNT within the granulomas [34]. This is similar, within limits, to observations with other fibrous inflammatory agents such as asbestos [33]. However, this qualitative observation has not been tested directly. It may be that larger SWCNT tangles have a greater inflammatory potential than smaller ones. If this is the case, however, the differences in magnitude are of an order that they did not present obvious histological differences in the current available studies.

Further uncertainty exists in the derivation of the SWCNT elimination rate based on two observations by Muller et al. [31] at an exposure rate different from that used as the toxicity threshold. Scaling the exposure to the projected risk threshold required an assumption of first-order kinetics. Regression of the one-dose observations suggests strongly that the elimination does follow first-order kinetics. However, it has not been repeated or demonstrated for other SWCNT exposure rates. It is currently an assumption and therefore represents an uncertainty in this derivation.

9.6 KNOWN TOXICITY OF NANOMATERIALS

The study of the toxicity of nanomaterials is in its infancy and the literature is growing rapidly. It is useful to examine the literature to date for the six types of nanomaterials that are the focus of this book to understand the types of effects that might occur. Table 9.1 offers a brief review of the literature.

As shown in Table 9.1, a number of studies have investigated the potential human health implications associated with exposure to nanomaterials. Although many of the results are very preliminary, there are indications that the six major engineered nanomaterials can elicit an oxidative stress response in certain biological test systems [37, 39, 54–56]. This is seen as measured indication of cell membrane damage,

TABLE 9.1
Effects of Nanomaterials or Nanoparticles on Mammalian Species

Species	Particle[a] Category	Size or Diameter	Exposure or Dose	Endpoint(s)	Effect(s)	Commentary	Ref.
Human mesothelioma and rodent fibroblast cell lines	Nanoparticle TiO_2	8 nm	0, 3.75, 7.5, and 15 μg/ml for 6-day periods and 0, 7.5, 15, and 30 μg/mL for 3 days exposure	Cytotoxicity	Weak cytotoxic effects	Both cell lines showed less response after 6 days of exposure compared to 3 days of exposure. This might be due to initial stress of the nanoparticles, and then detoxification of the particles and cell culture viability recovers.	[35]
Rats	TiO_2 particles Nanoscale TiO_2 rods Nanoscale TiO_2 dots	300 nm (rutile type) 200 nm × 35 nm (anatase type)10 nm (anatase type)	1 or 5 mg/kg intratracheally instilled in phosphate-buffered saline; evaluated at 24 hr, 1 week, 1 month, and 3 months post instillation	Oxidative stress/ cytotoxicity; lung histopathology	TiO_2 particles, dots, and rods caused transient inflammatory and cytotoxic effects observed at 24 hr post exposure, but effects were not sustained. Instilled quartz particles caused sustained dose-dependent inflammation as well as lung tissue damage consistent with pulmonary fibrosis development.	Results indicate that nanoscale particles might not be more cytotoxic to lung compared to larger-sized particles. In addition, results demonstrate that surface area might not be associated with pulmonary toxicity of nanoscale particles.	[36]
Syrian hamster embryo fibroblast cells	Ultrafine TiO_2	≤20 nm	Cells treated with 0.5, 1.0, 5, or 10 μg/cm2 for 12, 24, 48, 66, or 72 hr	Oxidative stress/ cytotoxicity	A significant dose (concentration)-dependent increase in micronuclei formation between 0.5 and 5.0 μg/cm2 (measurement of chromosomal change); cell death (apoptosis) observed after 24, 48, and 72 hr.	Results support the mechanism of cell death from exposure to nanoparticles: particles react with cell membranes, in turn generate reactive oxygen species (ROS); the oxidative stress leads to cell toxicity.	[37]

Model	Material	Size	Dose/Exposure	Endpoint	Results	Interpretation	Ref.
Mice (C57B1/6)	TiO_2	2–5 nm	0.77 or 7.22 mg/m³ (acute exposure, 4 hr); 8.8 mg/m³ (subacute exposure, 4 hr/day for 10 days)	Oxidative stress/cytotoxicity; lung histopathology	Acute exposure: at the high concentration, BAL fluid significantly increased; other parameters did not show inflammation. Subacute exposure: alveolar macrophages elevated in mice necropsied at weeks 0, 1, and 2 post exposure, not elevated at week 3 post exposure.	Minimal inflammatory response likely reflects a surface area threshold; anything below this threshold will cause little or no inflammatory response.	[38]
Mouse microglia cells	Nanosize TiO_2 (Degussa P25)	826–2368 nm	5–120 ppm for 6 or 18 hr	Oxidative stress	Significant release of ROS occurred at 60 min post exposure at concentrations ≥20 ppm; cell death not observed at all concentrations	Results demonstrate that TiO_2 can stimulate microglia to produce ROS; however, microglia remained viable. Further study to understand if the ROS translates into neural damage *in situ*.	[39]
Human red blood cells	TiO_2, anatase	0.02–0.03 m	5 µg/mL; incubated 4–24 hours	Red blood cells	TiO_2 aggregates with a diameter ≤0.2 µm were taken up by red blood cells; larger aggregates were stuck to surface of cell membrane.	Results suggest that nanoparticles may penetrate red blood cells by a mechanism other than phagocytosis and endocytosis	[40]
Mouse microglia; rat dopaminergic neurons; and embryonic rat striatum	TiO_2	Diameter of aggregates: 800 to 1900 nm (at 30 min); 770 nm (at 2 hr)	2.5–120 ppm	Neurotoxicity of nerve cells (microglia, neuron)	Cytotoxicity reported for microglia and striatum; TiO_2 did not cause toxicity to the dopaminergic neurons.	Results suggest that the neurotoxicity of TiO_2 is mediated through microglia-generated ROS.	[41]

TABLE 9.1 (CONTINUED)

Effects of Nanomaterials or Nanoparticles on Mammalian Species

Species	Particle[a] Category	Size or Diameter	Exposure or Dose	Endpoint(s)	Effect(s)	Commentary	Ref.
Mouse spermatogonial stem cell line	Nanoscale silver	15 nm	5, 10, 25, 50, and 10 µg/mL	Cytotoxicity; mitochondrial function, cell morphology, cell membrane leakage, and cell death	Silver nanoparticles were the most cytotoxic of the compounds tested. Cytotoxic effects were dose-dependent.	The spermatogonial cell line was chosen for this study to evaluate the toxicity of nanoparticles on the male germline. Results of this study suggest that the cell line is a good model.	[42]
Human epithelial cells	Nanoparticle carbon black, Fine carbon black, Titanium dioxide, Nanoparticle TiO$_2$	14.3 nm 260 nm 250 nm 29 nm	31.25–200 µg/mL	Cytotoxicity	Inflammatory response	The highly toxic nature and reactive surface chemistry of the carbon black nanoparticles very likely induced the type II cell line to release pro-inflammatory mediators that can potentially induce migration of macrophages.	[43]

Mice (LDLR/KO)	Carbon black	120.7 nm	Endotracheal dispersion of 1 mg per animal per week for 10 weeks or intratracheal dispersion of air and 1 mg per animal per week for 10 weeks. Diets were also controlled for 0 or 0.51% cholesterol. Acute study performed: animals fed 0.51% cholesterol diet for 3 days and a single 1 mg/animal dose of carbon black.	Aorta/circulatory system	Aortic lipid-rich lesions were reported in mice receiving the 0.51% diet with and without carbon black exposure. No lesions in mice receiving the 0.0% diet. Greatest amount of lesions were reported in 0.51% group with carbon black exposure.	Results indicate that respiratory exposure to carbon black might accelerate the development of atherosclerosis and be associated with cardiovascular adverse effects. [44]
Rat aortic smooth muscle cells	SWCNT	10–15 nm	0.0–0.1 mg/mL added to cells and incubated for periods of 1, 2.5, and 3.5 days	Cytotoxicity	Unfiltered SWCNT media: cell growth not affected after 1 day; decrease in cell growth at 2.5 days for concentrations from 0–0.05 mg/mL; filtered SWCNT (removal of SWCNT aggregates): increase in cell number for concentrations 0–0.05 mg/mL; growth inhibition at 0.1 mg/mL dose for both filtered and unfiltered SWCNT.	Results indicate that aggregates affect cell growth, but not solely responsible for the cytotoxicity. [45]

TABLE 9.1 (CONTINUED)
Effects of Nanomaterials or Nanoparticles on Mammalian Species

Species	Particle[a] Category	Size or Diameter	Exposure or Dose	Endpoint(s)	Effect(s)	Commentary	Ref.
Human leukemic cells	Single- and multi-walled carbon nanotubes; 3 different samples by different synthesis: Sample 1 — MWCNTs (synthesized by an electric discharge) Sample 2 — 50% MW CNTs + 30% SWCNTs Sample 3 — MWCNTs (purchased)	Sample 1: 10–50 nm Sample 2: 10–40 nm Sample 3: 110–170 nm	25 μg/ml of each of the three samples	Oxidative stress/ cytotoxicity	Cytotoxic effects not observed; cell growth rate reduced	With the lack of cytotoxicity, the decrease in cell growth might be a result of the carbon nanotubes affecting the cell cycle directly.	[46]
Rat (in vivo) peritoneal macrophages (in vitro)	Multi-walled carbon nanotubes (CNT) and ground CNT	9.7 nm (CNT) 11.3 nm (ground CNT)	0.5, 2, or 5 mg/animal intratracheally instilled (in vivo)	Pulmonary toxicity; inflammatory response; fibrotic response	In vivo results: inflammatory response and dose-dependent pulmonary fibrosis for both CNT and ground CNT, effects more pronounced with ground CNT. In vitro results: ground CNT increased macrophage production.	Results indicate that multi-wall carbon nanotubes are capable of eliciting inflammatory and fibrotic response in lungs	[31]

Model	Material	Size	Dose/Exposure	Endpoint	Results	Conclusions	Ref.
Mice	Single-walled carbon nanotubes	1–4 nm	Acute: 10–40 μg/mouse, single intrapharyngeal instillation (sacrificed at 1, 7, 28, and 56 days post exposure) Chronic: 20 μg/mouse, pharyngeal aspiration, once every other week for 8 weeks (mice in this group are bred with elevated cholesterol levels and were fed a high-fat diet).	Aortic mitochondria (oxidative stress assays)	Aortic mitochondrial DNA damage at 7, 28, and 60 days post exposure. Increase in atherosclerosis plaque formation reported in chronically treated mice.	Results indicate that respiratory exposure to SWCNTs might accelerate the development of atherosclerosis and be associated with cardiovascular adverse effects.	[47]
Guinea pig alveolar macrophages	SWCNTs, MWCNT10, and fullerene (C60)	1.4 nm (SWCNT) 10–20 nm (MWCNT10) C60 — not provided	SWCNT and C60: 0, 1.41, 2.82, 5.65, 11.30, 28.20, 56.50, 113.00, and 226.0 μg/cm² MWCNT10: 0, 1.41, 2.82, 5.65, 11.30, and 22.60 μg/cm²	Cytotoxicity	SWCNT: high cytotoxicity at lowest dose; for MWCNT10, cytotoxic effects but at the highest dose. C60 did not induce cytotoxicity at any concentration.	Results suggest that toxicity of nanomaterials increases with an increase in surface area.	[34]
Mice (B6C3F$_1$)	SWCNT, Carbon black	1 nm	Single dose of 0, 0.1, or 0.5 mg/mouse intratracheally instilled; euthanized 7 and 90 days post treatment	Cytotoxicity/lung	SWCNT: Dose-dependent epithelioid granulomas, interstitial inflammation, peribronchial inflammation, necrosis. Lesions more persistent and pronounced in 90-day group. Carbon black: no lung adverse effects.	Results suggest that toxicity of nanomaterials increases with an increase in surface area.	[23]

TABLE 9.1 (CONTINUED)

Effects of Nanomaterials or Nanoparticles on Mammalian Species

Species	Particle[a] Category	Size or Diameter	Exposure or Dose	Endpoint(s)	Effect(s)	Commentary	Ref.
Rat	Single-walled carbon nanotube (SWCNT)	Diameters < 2 nm, with lengths ranging from 0.5 to 40 μm and a purity > 90%.	Oropharyngeal aspiration of 2 mg/kg-bw. Evaluated (bronchoalveolar lavage) 1 and 21 days post exposure	Lung–lung histopathology, fibrogenic potential, cell proliferation, and growth factor mRNAs. Exposure biomarker	SWCNT did not cause lung inflammation, but induced the formation of small, focal interstitial fibrotic lesions in the alveolar region of rat lungs.	Of greatest interest — unique intercellular carbon structures composed of SWCNT-bridged lung macrophages. These bridges offer an easily identifiable exposure biomarker.	[48]
C57Bl/6 Mice	SWCNT	1–4 nm	0, 10, 20, or 40 μg/mouse; pharyngeal aspiration. Animals euthanized at 1, 3, 7, 28, and 60 days post exposure	Oxidative stress/cytotoxicity lung	Acute inflammation with early onset, progressive fibrosis and granulomas. Dose-dependent increase in oxidative stress biomarkers. Functional respiratory deficiencies and decreased bacterial clearance also observed.	Results support *in vitro* studies.	[24]
Human fibroblasts	SWCNT MWCNT Carbon black Activated carbon Carbon graphite	2 nm 200 nm 25 nm 50 nm 500 nm	0.8, 1.61, 3.125, 6.25, 12.5, 25, 50, and 100 μg/mL for 1 to 5 days	Cytotoxicity	Cellular apoptosis/necrosis. SWCNTs induced strongest cellular apoptosis/necrosis. Refined SWCNTs are more toxic than unrefined counterpart.	Results suggest that toxicity of nanomaterials increases with increase in surface area.	[26]

Species	Size	Material	Dose/Exposure	Endpoint	Results	Comments	Ref.
Rats	30 nm	SWCNT	1 or 5 mg/kg intratracheally instilled; evaluated 24 hr, 1 week, 1 month, and 3 months post exposure	Cytotoxicity/lung	Transient inflammatory and cell injury effects, non-dose-dependent series of multi-focal granulomas.	Physiological relevance of these findings should be determined by conducting an inhalation toxicity study.	[25]
Mouse L929 fibrosarcoma Rat C6 glioma U251 human glioma	100 nm <5 nm	C60 fullerence $C60(OH)_n$ — polyhydroxylated fullerene	1 or 1000 (g/mL)	Oxidative stress/cytotoxicity	Cytotoxic action reached maximum after 6 hr with C60; minimal cytotoxicity for the same time period for $C60(OH)_n$; Reactive oxygen species (ROS) not produced in cells treated with $C60(OH)_n$; rapid increase of ROS in cells treated with C60.	Results demonstrate that C60 is at least three orders of magnitude more toxic than $C60(OH)_n$.	[49]
Rats	160 ± 50 nm	C60 fullerenes	0.2, 0.4, 1.5, or 3.0 mg/kg intratracheally instilled; lung tissue evaluated 1 day, 1 week, 1 month, and 3 months post instillation	Oxidant and glutathione endpoints; bronchoalveolar lavage (BAL) fluid biomarkers; lung tissue	Transient inflammatory and cell injury at 1 day post exposure, not different from controls at other post exposure times; BAL biomarkers increased in 1.5 and 3.0 mg/kg dose groups at 1 day and 3 months post exposure; no adverse lung tissue effects at 3 months post exposure at any dose.	Results not consistent with results reported for in vitro studies; such findings highlight the difficulty in extrapolating in vitro effects to in vivo effects.	[50]

TABLE 9.1 (CONTINUED)
Effects of Nanomaterials or Nanoparticles on Mammalian Species

Species	Particle[a] Category	Size or Diameter	Exposure or Dose	Endpoint(s)	Effect(s)	Commentary	Ref.
Human umbilical vein endothelial cells	Hydroxyl fullerene ($C_{60}(OH)_{24}$)	7.1 ± 2.4 nm	1–100 µg/mL for 24 hr (acute exposure) 1–10 µg/mL for up to 10 days (chronic exposure)	Cytotoxicity	Acute exposure: cytotoxic morphological effects; cell growth inhibited; fullerenes did not induce apoptosis chronic exposure: 1 µg/mL for 10 days had no effect on endothelial cell toxicity; at 10 µg/mL, caused cytotoxicity.	Because nanomaterials can move into the circulation, these materials might affect vascular cells, leading to vascular injury and possible atherosclerosis development. This study demonstrates that exposure to nanomaterials might lead to cardiovascular effects; however, *in vivo* studies are necessary to validate.	[51]
Human platelet (*in vitro*) Rat (*in vivo*)	C60 fullerenes Multi-walled nanotubes Single-walled nanotubes	Not available	0.2–300 µg/mL (human platelet) 50 µg/mL (rat)	Human platelet cells Rat vascular thrombosis (carotid artery)	Dose-dependent increase in platelet aggregation with single- and multi-walled nanotubes, not with C60 fullerenes. Single- and multi-walled nanotubes significantly increased the time and rate of carotid artery thrombosis; C60 fullerenes did not impact the development of thrombosis.	Results indicate that blood platelets might be activated (enhancing vascular thrombosis) by exposure to carbon nanoparticles *in vivo*. Second, it appears that nanospheres (e.g., C60 fullerenes) do not promote cell-cell contact (like platelet aggregation), whereas nanotubes seem to cause aggregation.	[52]

| Human lymphocytes | EthOH/nC60 | 121.8 nm | 4.20 mg/L | Genotoxicity | Genotoxicity for both types of suspensions with a strong correlation between the genotoxic response and nC60 concentration. | Results demonstrated a strong relationship between C60 and DNA damage. Mechanism of how the damage occurs requires further study. | [53] |
| | aqu/nC60 | 211.8 nm | 0.23 mg/L | | | | |

^a Acronyms: SWCNT: single-walled carbon nanotubules; MWCNT: multiwalled carbon nanotubules; CB: carbon black; TiO$_2$: titanium dioxide.

inflammation, DNA damage, and apoptosis (cell death) [49]. The mechanism for this oxidative stress response is neither always clear nor consistent. In some systems, the oxidative stress appears to be due to direct production of reactive oxygen species (ROS) by the nanoparticles [39, 55]. In others, it appears the damage may be mediated through a direct inflammation response to the nanoparticles. The studies to date also suggest that toxicity increases directly with the surface area per unit mass of the nanomaterial [23, 26, 34, 38, 45, 57–59].

While the lung is the primary target of nanomaterials [31, 54, 57], studies show that the nanoparticles can enter circulation and migrate to various organs and tissues, where they can accumulate and damage organ systems that are sensitive to oxidative stress [40, 44, 57]. Structural impacts such as vascular thrombosis, accelerated development of atherosclerosis, other adverse cardiovascular effects, and neurodegeneration are common toxic responses to oxidative toxicants. Because the study of nanotoxicity is in its infancy, many of the studies to date are *in vitro* studies and therefore do not consider pharmacokinetic factors. Much more study is required in evaluating the toxicity of nanomaterials, particularly *in vivo* systems, before the true toxicological nature of these materials can be discerned. [24, 25, 39, 50].

9.7 CONCLUSIONS

The health risk assessment of nanomaterials requires certain considerations not common with typical toxicants. First, the characterization of nanomaterials will vary significantly based on whether the material is quantified on a mass or molar basis because of the potential for high variability for the same chemical based on the size distribution of the particle considered. If a nanomaterial will be assessed based on its activity as a nanoparticle, then an approach similar to that used for the assessment of exposure to fibers should be used, rather than the typical approach that characterizes risk on a per-mass basis. Second, nanomaterials' physical size introduces novel factors related to exposure and absorption based on steric factors that affect both dosing rates and *in vivo* kinetics. Consequently, variations in the sensitivity of the population become of paramount importance to predicting adverse response. The third, and likely the most troublesome, consideration is that the types of toxicological response will likely vary between molecular and nano-sized forms. Nanomaterials are less likely to induce systemic effects and more likely to act as contact toxicants. Furthermore, the size of the material's particles makes it likely that the physiological responses will be of an immune or inflammatory nature. These adverse effects based on defensive mechanisms typically do not follow dose-response but rather threshold kinetics, and also demonstrate high variability within the population, depending on individual sensitivities.

9.8 LIST OF SYMBOLS

A	Surface area
A_Y	Surface area at Ydt
A_{GI}	Cellular surface presented to the gastrointestinal (GI) tract
A_{CC}	Cellular surface presented to the central compartment

A_{mem} Membrane surface area
$[C]_{GI}$ Solute concentration within the GI tract
$[C]_{IC}$ Solute concentration in the GI epithelium cell
$[C]_{CC}$ Solute concentration within the central compartment
$[C_0]$ Initial concentration
D Diffusion coefficient
D_{mem} Diffusion coefficient of the membrane
d_{mem} Diffusion distance or membrane thickness
CS_Y Cross-sectional area of the airway at point y
dc Concentration gradient
dx Distance across the concentration gradient
dn_1/dt Rate of solute flux from GI tract to GI epithelial cell
dn_2/dt Rate of solute flux from GI epithelial cell to the central compartment
dC/dx Concentration gradient
dy Infinitesimal of the change in longitudinal position within the airway
J Flux rate
J_{mem} Flux rate across the membrane boundary
k Boltzmann constant
k_{ow} Octanol-water partition coefficient
k_{mem} Partition coefficient of a solute between an aqueous solution/suspension and a membrane solution/suspension
n Number of particles
P Permeability
P^{v-o} Theoretical zero-volume permeability
r_p Radius of the nanoparticle
r_a Radius
T Absolute temperature
$v-$ Air flow velocity
V Molecular volume
y Hypothetical point
η Viscosity of the aerosol

REFERENCES

1. McCash, E.M. 2001. *Surface Chemistry.* Oxford: Oxford University Press.
2. Freeman, H.J. and A.B.R. Thomson. 2000. Small intestine. In *First Principals of Gastroenterology: The Basis of Disease and an Approach to Management,* 4th edition, Eds. A.B.R. Thompson and E.A. Shaffer, p. 175–257. Edmonton, AB: Astra.
3. Lieb, W.R. and W.D. Stein. 1986. Non-Stokesian nature of transverse diffusion within human red cell membranes. *J. Membrane Biol.,* 92:111–119.
4. Sherrill, B.C. and J.M. Dietschy. 1978. Characterization of the sinusoidal transport process responsible for uptake of chylomicrons by the liver. *J. Biolog. Chem..* 253:1859–1867.
5. Jani, P., G.W. Halbert, J. Langridge, and A.T. Florence. 1990. Nanoparticle uptake by rat gastrointestinal mucosa: quantitation and particle size dependency. *J. Pharm. Pharmacol.,* 42:821–826.

6. Lomer, M.C., R.P. Thompson, and J.J. Powell. 2002. Fine and ultrafine particles of the diet: Influence on the mucosal immune response and association with Crohn's disease. *Proc. Nutr. Soc.*, 61:123–130.

7. Wanga, B., W. Fenga, T. Wangc, et al. 2006. Acute toxicity of nano- and micro-scale zinc powder in healthy adult mice. *Toxicol. Lett.*, 161:115–123.

8. Ballard, W.E. 1972. Some environmental problems of the metal spraying processes. *Ann. Occup. Hyg.*, 15:101–106.

9. Raynor, P.C., S. Kim, and M. Bhattacharya. 2005. Mist generation from metalworking fluids formulated using vegetable oils. *Ann. Occup. Hyg.*, 49:283–293.

10. Lademann, J., H. Weigmann, C. Rickmeyer, et al. 1999. Penetration of titanium oxide microparticles in a sunscreen formulation into the horny layer and the follicular orifice. *Skin Pharmacol. Appl. Skin Physiol.*, 12:247–256.

11. Hostynek, J.J. 2003. Factors determining percutaneous metal absorption. *Food Chem. Toxicol.*, 41:327–345.

12. Verma, D.D., S. Verma, G. Blume, and A. Fahr. 2003. Particle size of liposomes influences dermal delivery of substances to the skin. *Int. J. Pharm.*, 258:141–151.

13. Rouse, J.G., J. Yang, J.P. Ryman-Rasmussen, A.R. Barron, and N.A. Monteiro-Riviere. 2007. Effects of mechanical flexion on the penetration of fullerene amino acid-derivatized peptide nanoparticles through skin. *Nano Lett.*, 7:155–160.

14. Hoet, P.H.M., I. Bruske-Hohlfeld, and O.V. Salta. 2004. Nanoparticles — known and unknown health risks. *J. Nanobiotechnol.*, 2:12–25.

15. Menon, G.K. and P.M. Elias. 1997. Morphologic basis for a pore pathway in mammalian stratum corneum. *Skin Pharmacol.*, 10:235–246.

16. De Jalon, E.G., M.J. Blanco-Prieto, P. Ygartua, and S. Santoyo. 2001. PLGA microparticles: Possible vehicles for topical drug delivery. *Int. J. Pharm.*, 226:181–184.

17. Tinkle, S.S., J.M. Antonini, B.A. Rich, et al. 2003. Skin as a route of exposure and sensitization in chronic beryllium disease. *Environ. Health Perspect.*, 111:1201–1208.

18. Dorman, D.C., B.E. McManus, C.U. Parkinson, C.A. Manuel, A.M. McElveen, and J.I. Everitt. 2004. Nasal toxicity of manganese sulfate and manganese phosphate in young rats following subchronic (13-week) inhalation exposure. *Inhalation Toxicol.*, 16:481–488.

19. Schroeter, J.D., J.S. Kimbell, M.E. Andersen, and D.C. Dorman. 2006. Use of a pharmacokinetic-driven computational fluid dynamics model to predict nasal extraction of hydrogen sulfide in rats and humans. *Toxicol Sci.*, 94:359–367.

20. Hollinger, M.A. 1985. Respiratory pharmacology and toxicology. In *Saunders Monographs in Pharmacology and Therapeutics*. Philadelphia: W.B. Saunders Co.

21. Berthiaume, E.P., C. Medina, and J.A. Swanson. 1995. Molecular size-fractionation during endocytosis in macrophages. *J. Cell Biol.*, 129:989–998.

22. Lam, C.-W., J.T. James, R. McCluskey, S. Arepalli, and R.L. Hunter. 2006. A review of carbon nanotube toxicity and assessment of potential occupational and environmental health risk. *Crit. Rev. Toxicol.*, 36:189–217.

23. Lam, C.-W., J.T. James, R. McCluskey, and R.L. Hunter. 2004. Pulmonary toxicity of single-walled carbon nanotubes in mice 7 and 90 days after intratracheal instillation. *Toxicol. Sci.*, 77:126–134.

24. Shévedova, A.A., E.R. Kisin, R. Mercer, et al. 2005. Unusual inflammatory and fibrogenic pulmonary responses to single-walled carbon nanotubes in mice. *Am. J. Physiol. Lung Cell. Molec. Physiol.*, 289:L698–L698.

25. Warheit, D.B., B.R. Laurence, K.L. Reed, D.H. Roach, G.A.M. Reynolds, and T.R. Webb. 2004. Comparative pulmonary toxicity assessment of single-wall carbon nanotubes in rats. *Toxicol Lett.*, 77:117–125.

26. Tian, F., D. Cui, H. Schwarz, G.G. Estrada, and H. Kobayashi. 2006. Cytotoxicity of single-walled carbon nanotubes on human fibroblasts. *Toxicol. In Vitro*, 20:1202–1212.

27. Witschi, H.R. and J.A. Last. 1996. Toxic responses of the respiratory system. In *Casarett and Doull's Toxicology*, 5th edition, p. 443–462.
28. Shusterman, D. 2002. Review of the upper airway, including olfaction, as mediator of symptoms. *Environ. Health Perspect.*, 110:649–653.
29. Penn, A., G. Murphy, S. Barker, W. Henk, and L. Penn. 2005. Combustion-derived ultrafine particles transport organic toxicants to target respiratory cells. *Environ. Health Perspect.*, 113(8):956–963.
30. Linstedt, S.L. and P.J. Schaffer. 2002. Use of allometry in predicting anatomical and physiological parameters of mammals. *Lab. Animals,* 36:1–19.
31. Muller, J., F. Huaux, and N. Moreau. 2005. Respiratory toxicity of multi-walled carbon nanotubes. *Toxicol. Appl. Pharmacol.*, 207:221–231.
32. U.S. Environmental Protection Agency. 1997. *Exposure Factors Handbook*. Office of Research and Development. EPA/600/P-95/00Fa.
33. Selikoff, I.J. 1990. Historical developments and perspectives in inorganic fiber toxicity in man. *Environ. Health Perspec.*, 88:269–276.
34. Jia, G., H. Wang, L. Yan, et al. 2005. Cytotoxicity of carbon nanomaterials: Single-wall nanotubes, multi-wall nanotubes, and fullerene. *Environ. Sci. Technol.*, 39:1378–1383.
35. Brunner, T.J., P. Wick, P. Manser, et al. 2006. *In vitro* cytotoxicity of oxide nanoparticles: comparison to asbestos, silica, and the effects of particle solubility. *Environ. Sci. Technol.*, 40:4374–4381.
36. Warheit, D.B., T.R. Webb, S.M. Sayes, V.L. Colvin, and K.L. Reed. 2006. Pulmonary instillation studies with nanoscale TiO_2 rods and dots in rats: Toxicity is not dependent upon particle size and surface area. *Toxicol. Sci.*, 91(1):227–236.
37. Rahman, Q., M. Lohani, E. Dopp, et al. 2002. Evidence that ultrafine titanium dioxide induces micronuclei and apoptosis in Syrian hamster embryo fibroblasts. *Environ. Health Perspect.*, 110(8):797–800.
38. Grassian, V.H., P.T. O'Shaughnessy, A. Adamc, J.M. Pettibone, and P.S. Thorne. 2007. Inhalation exposure study of titanium dioxide nanoparticles with a primary particle size of 2 to 5 nm. *Environ. Health Perspect.*, 115(3):397–402.
39. Long, T.C., N. Saleh, R.D. Tilton, G.V. Lowry, and B. Veronesi. 2006. Titanium dioxide (P25) produces reactive oxygen species in immortalized brain microglia (BV2): Implications for nanoparticle neurotoxicity. *Environ. Sci. Technol.*, 40(14):4346–4352.
40. Rothen-Rutishauser, B.M., S. Schurch, B. Haenni, N. Kapp, and P. Gehr. 2006. Interaction of fine particles and nanoparticles with red blood cells visualized with advanced microscopic techniques. *Environ. Sci. Technol.*, 40:4353–4359.
41. Long, T.C., J. Tajuba, P. Sama, et al. 2007. Nanosize titanium dioxide stimulates reactive oxygen species in brain microglia and damages neurons in vitro. *Environ. Health Perspect.*, 115(11):1631–1637.
42. Braydich-Stolle, L., S. Hussain, J.J. Schlager, and M.-C. Hofmann. 2005. *In vitro* cytoxicity of nanoparticles in mammalian germline stem cells. *Toxicolog. Sci.,* 88(2):412–419.
43. Barlow, P.G., A. Clouter-Baker, K. Donaldson, J. MacCallum, and V. Stone. 2005. Carbon black nanoparticles induce type II epithelial cells to release chemotaxins for alveolar macrophages. *Particle Fibre Toxicol.*, 2:11.
44. Niwa, Y., Y. Hiura, T. Murayama, M. Yokode, and N. Iwai. 2007. Nano-sized carbon black exposure exacerbates atherosclerosis in LDL-receptor knockout mice. *Circ. J.,* 71:1157–1161.
45. Raja, P.M.V., J. Connolley, G.P. Ganesan, et al. 2007. Impact of carbon nanotube exposure, dosage and aggregation on smooth muscle cells. *Toxicol. Lett.*, 169(1):51–63.
46. De Nicola, M., D.M. Gattia, S. Bellucci, et al. 2007. Effect of different carbon nanotubes on cell viability and proliferation. *J. Phys. Condens. Matter,* 19:1–7.

47. Li, Z., T. Hulderman, R. Salmen, et al. 2007. Cardiovascular effects of pulmonary exposure to single-wall carbon nanotubes. *Environ. Health Perspect.,* 115(3):377–382.
48. Magnum, J.B., E.A. Turpin, A. Antao-Menezes, M.F. Cesta, E. Bermudez, and J.C. Bonner. 2006. Single-walled carbon nanotube (SWCNT)-induced interstitial fibrosis in the lungs of rats is associated with increased levels of PDGF mRNA and the formation of unique intercellular carbon structures that bridge alveolar macrophages *in situ. Particle Fibre Toxicol.,* 3:15.
49. Isakovic, A., Z. Markovic, B. Todorovic-Markovic, et al. 2006. Distinct cytotoxic mechanisms of pristine versus hydroxylated fullerene. *Toxicolog. Sci.,* 91(1):173–183.
50. Sayes, C.M., A.A. Marchione, K.L. Reed, and D.B. Warheit. 2007. Comparative pulmonary toxicity assessment of C_{60} water suspensions in rats: Few differences in fullerene toxicity *in vivo* in contrast to *in vitro* profiles. *Nano Lett.,* 7(8):2399–2406.
51. Yamawaki, H. and N. Iwai. 2006. Cytotoxicity of water-soluble fullerene in vascular endothelial cells. *Am. J. Physiol. Cell Physiol.,* 290:C1495–C1502.
52. Radomski, A., P. Jurasz, D. Alonso-Escolano, et al. 2005. Nanoparticle-induced platelet aggregation and vascular thrombosis. *Br. J. Pharmacol.,* 146:882–893.
53. Dahawan, A., J.S. Taurozzi, A.K. Pandey, et al. 2006. Stable colloidal dispersions of C60 fullerenes in water: Evidence for genotoxicity. *Environ. Sci. Technol.,* 40:7394–7401.
54. Helland, A., P. Wick, A. Koehler, K. Schmid, and C. Som. 2007. Reviewing the environmental and human health knowledge base of carbon nanotubes. *Environ. Health Perspect.,* 115(8):1125–1131.
55. Limbach, L.K., P. Wick, P. Manser, R.N. Grass, A. Bruinink, and W.J. Stark. 2007. Exposure of engineered nanoparticles to human lung epithelial cells: Influence of chemical composition and catalytic activity on oxidative stress. *Environ. Sci. Technol.,* 41:4158–4163.
56. Zhang, L.W., L. Zeng, A.R. Barron, and N.A. Monteiro-Riviere. 2007. Biological interactions of functionalized single-wall carbon nanotubes in human epidermal keratinocytes. *Int. J. Toxicol.,* 26(2):103–113.
57. Medina, C., M.J. Santos-Martinez, A. Radomski, O.I. Corrigan, and M.W. Radomski. 2007. Nanoparticles: Pharmacological and toxicological significance. *Br. J. Pharmacol.,* 150:552–558.
58. Magrez, A., S. Kasas, V. Salicio, et al. 2006. Cellular toxicity of carbon-based nanomaterials. *Nano Lett.,* 6(6):1121–1125.
59. Singh, S., T. Shi, R. Duffin, et al. 2007. Endocytosis, oxidative stress and IL-8 expression in human lung epithelial cells upon treatment with fine and ultrafine TiO_2: Role of the specific surface area and of surface methylation of the particles. *Toxicol. Appl. Pharmacol.,* 222(2):141–151.

10 Nanoparticle Use in Pollution Control

Kathleen Sellers
ARCADIS U.S., Inc.

CONTENTS

Given their high reactivity, it comes as no surprise that some nanoparticles find use in environmental remediation and related applications such as wastewater treatment and pollution prevention. This use leads to an apparent paradox: in an effort to improve conditions in the environment, materials with uncertain health and environmental effects may be released into the environment. One authority [1] notably said about this practice:

> "We recommend that the use of free (that is, not fixed in a matrix) manufactured nanoparticles in environmental applications such as remediation be prohibited until appropriate research has been undertaken and it can be demonstrated that the potential benefits outweigh the potential risks."
>
> — **The Royal Society and the Royal Academy of Engineering, 2004**

This chapter examines the use of engineered nanomaterials in environmental remediation and related applications such as wastewater treatment. It explores the apparent paradox in doing so and whether, since the British Royal Society and Royal Academy of Engineering issued their caution in 2004, we have learned enough to demonstrate that the benefits outweigh the risks. Nano zero-valent iron (nZVI) is perhaps the most widely used nanomaterial in environmental remediation and is described in some detail below. This chapter also includes information on other nanomaterials under development or currently in use to treat groundwater or wastewater, or in other pollution-control applications.

The information presented in this chapter originated from a combination of peer-reviewed literature, "gray" literature such as conference proceedings, and information from vendors. Readers should consult the references section for the basis for information presented in this chapter. Due to the rapid developments in the field, and at times to the need to protect confidential business information, supporting data for some of the referenced information are not always available. Mention of a specific product or brand name does not constitute endorsement.

10.1   ZERO-VALENT IRON (ZVI)

Zero-valent iron (ZVI) is used to treat recalcitrant and toxic contaminants such as chlorinated hydrocarbons and chromium in groundwater [2]. The initial applications used granular iron, alone or mixed with sand to make "magic sand," to treat extracted groundwater. Later, engineers installed flow-through ZVI cells in the ground, using slurry walls or sheet piling to direct the flow of groundwater through the treatment cells. However, these walls were expensive and sometimes difficult to construct, and often incurred long-term costs for maintenance and monitoring. Injectable forms of ZVI, most recently nano zero-valent iron (nZVI) and its variations, were developed to surmount these problems. In these applications, nanoscale iron particles are injected directly into an aquifer to effect treatment *in situ*. As described below, nZVI is commercially available and has been used on more than 30 sites as of this writing.

Zero-valent iron (Fe^0) enters oxidation-reduction (redox) reactions that degrade certain contaminants, particularly chlorinated hydrocarbons such as trichloroethylene (TCE) and tetrachloroethylene. ZVI also has been used to treat arsenic and certain metals [3]. In the presence of oxygen, nZVI can oxidize organic compounds such as phenol [4]. Much of the discussion in this chapter pertains to the treatment of chlorinated hydrocarbons because of the prevalence of those contaminants and resulting focus on their remediation using nZVI.

Reductive dehalogenation of TCE generally occurs as follows [5]:

$$Fe^0 \rightarrow Fe^{2+} + 2e^-$$

$$3Fe^0 + 4H_2O \rightarrow Fe_3O_4 + 8H^+ + 8e^- \tag{10.1}$$

$$TCE + n{\cdot}e^- + (n\text{-}3){\cdot}H^+ \rightarrow Products + 3Cl^-$$

$$H^+ + e^- \rightarrow H^{\cdot} \rightarrow \tfrac{1}{2} H_2\uparrow$$

where the value of n depends on the products formed. As indicated by these half-reactions, nZVI can be oxidized to ferrous iron or to Fe_3O_4 (magnetite); the latter is more thermodynamically favored above pH 6.1. As reaction proceeds, ZVI particles can become coated with a shell of oxidized iron (i.e., Fe_3O_4 and Fe_2O_3). This coating can eventually reduce the reactivity of (or "passivate") the nZVI particles [4, 5]. Passivation can begin immediately upon manufacture, depending on how the material is stored and shipped; the oxidation reaction continues after environmental application.

The efficiency of treatment depends on the rate of TCE dechlorination relative to nonspecific corrosion of the nZVI to yield H_2. In one study with granular ZVI, the latter reaction consumed over 80% of Fe^0 [5]. The solution pH and the Fe^0 content of the particles may affect the balance between nonspecific corrosion and reduction of TCE.

The effectiveness of *in situ* treatment using nZVI also depends on the characteristics of the aquifer. The pattern and rate of groundwater flow affect the distribution of nZVI. The geochemical characteristics of the groundwater — including pH, relative degree of oxygenation, and presence of naturally occurring minerals — also affect the reactivity and distribution of nZVI.

The remainder of this section provides more information on nZVI reagents, describing the size of nZVI particles and the effects of particle size, other constituents of nZVI reagents, and factors that affect the mobility of nZVI in the subsurface. It describes how sites are remediated with nZVI and presents examples. Finally, it discusses information on the potential risks from using nZVI and some of the resulting risk management decisions.

10.1.1 FORMS OF NZVI

nZVI can be manufactured using different processes that convey different properties to the material. These properties include particle size (and size distribution), surface area, and presence of trace constituents. Reagents for environmental remediation often contain materials other than iron to enhance the mobility or reactivity of nZVI.

In general, four processes are used to manufacture nZVI [7–9]:

1. Heat iron pentacarbonyl
2. Ferric chloride + sodium borohydride *
3. Iron oxides + hydrogen (high temperatures) *
4. Ball mill iron filings to nano-sized particles

The processes marked with an asterisk (*) are currently used in commercial production. Researchers have modified nZVI particles to increase their mobility and/or reactivity. Coating the nZVI particles can limit agglomeration and deposition, and enhance their dispersion. These particle treatments include emulsified nZVI, polymers, surfactants, and polyelectrolytes [10].

Bimetallic nanoscale particles (BNPs) have a core of nZVI with a trace coating of a catalyst such as palladium, silver, or platinum [11]. This catalyst enhances reduction reactions. PARS Environmental markets a BNP developed at Penn State University. This BNP contains 99.9 wt% iron and 0.1 wt% palladium and polymer support. The polymer is not toxic; the U.S. Food and Drug Administration has approved the use of the polymer as a food additive. The polymer limits the ability of the nZVI particles to agglomerate and adhere to soils. Case studies presented later in this chapter describe the use of this BNP to degrade chlorinated solvents in groundwater.

10.1.2 Particle Characteristics

The particle size and other characteristics of nZVI depend, in part, on the method of synthesis [7–9]. Two studies have measured the actual particle sizes in commercially available nZVI. These studies also provided information on the surface area of the particles and their elemental composition. The particle size and resultant surface area affect the mobility and reactivity of the iron nanoparticles.

Nurmi et al. [12] tested nZVI samples from Toda Kogyo Corporation's RNIP-10DS product. The manufacturer indicates that the nZVI particles are approximately 70 nm in diameter and have a surface area of 29 square meters per gram (m^2/g). RNIP-10DS is produced by reacting iron oxides (goethite and hematite) with hydrogen at temperatures between 200 and 600°C. The resulting iron particles contain Fe^0 and Fe_3O_4 (in total, approximately 70 to 30% iron and 30 to 70% oxide) based on x-ray diffraction analysis (XRD). X-ray photoelectron spectroscopy indicated that the particles also contained trace amounts of S, Na, and Ca. Nurmi et al. [12] used transmission electron microscopy (TEM) to examine the particle geometry. The nZVI consisted of aggregates of small, irregularly shaped particles of a nearly crystal Fe^0 core with an outer shell of polycrystalline iron oxide. TEM indicated that the average particle size in RNIP-10DS, as received, was 38 nm and the average surface area 25 m^2/g.

In another study, the Polyflon division of Crane Co. commissioned Lehigh University and the Whitman Companies Inc., through ARCADIS, to characterize the iron particles in four samples of PolyMetallix™ nZVI [13]. The method for synthesizing PolyMetallix™ nZVI was not specified, other than to indicate that Polyflon had treated some of the product samples via physical size reduction and/or the addition of a dispersing agent after the initial synthesis. Three of the samples were analyzed within approximately 2 weeks of manufacture. The fourth sample was analyzed more than 4 months after manufacture. In general, the age of the sample affected the particle size more than did the post-synthesis treatments. TEM showed that the nZVI comprised generally spherical particle clusters, with some of the clusters agglomerated. The older sample showed greater agglomeration. The mean particle size for the samples analyzed within 2 weeks of manufacture ranged from 66.0 to 68.5 nm; the mean nZVI size for the older sample was 186.8 nm. Each of these means represented a particle size distribution. For example, the particles in the aged sample ranged in size from 37.7 to 512.7 nm, with most of the particles between 125 and 300 nm. The study concluded, in part, that:

> "While the PSD [particle size distribution] is an important quality assurance and quality control parameter, it alone is not a sufficient indicator of nZVI reactivity or efficacy in a given remediation scenario. It is important to emphasize that nZVI in general are highly reactive materials and, as such, their surface and intrinsic properties change rapidly over time from the time of manufacture."

10.1.3 EFFECTS OF PARTICLE SIZE

How does the particle size relate to the reactivity of nZVI? As described in Chapter 2, nanoparticles may behave differently than their bulk counterparts due to the increased relative surface area per unit mass and/or the influence of quantum effects. As discussed below, the typical particle sizes of nZVI and experience with granular ZVI provide insight into why nZVI can be so effective.

For a metal such as iron, quantum effects on physical and chemical properties are negligible above a particle size of approximately 5 nm. (For metal oxides, which have a lower electron density, quantum effects may become evident at particle sizes between 10 and 150 nm [12].) Therefore, given the typical particle sizes of commercially available nZVI, quantum effects are probably negligible. The effectiveness of nZVI must relate, then, to particle size rather than to quantum behavior.

Previous work with granular (not nano) ZVI showed that the rate of reductive dehalogenation is relatively independent of contaminant concentration and depends strongly on the surface area of the iron catalyst [2]. The smaller the particle, the higher the percentage of the total number of atoms on the surface of the particle, and thus the higher the reactivity. A comparison of degradation rates for carbon tetrachloride treated by granular ZVI and nZVI showed that the higher reaction rate with nZVI resulted from the high surface area, not from a greater relative abundance of reactive sites on the surface of nZVI or the greater intrinsic reactivity of surface sites on nZVI [6, 12]. Some data suggest that reaction with nZVI can generate different products than reaction with granular ZVI, although the mechanisms causing this apparent difference are not yet understood [12].

Over time, agglomeration increases the effective particle size. This has been observed, as described above, in aged reagent samples. Increases in particle sizes can limit the mobility of the nZVI because larger particles cannot remain suspended in and transported by the groundwater. Consideration of the primary physical forces acting on nZVI particles suspended in water, as discussed in Section 6.2.1 and shown in Figure 6.4, suggests that less than half the particles above 80 nm in size will remain in stable suspension. Phenrat et al. [81] studied the agglomeration of nZVI in laboratory experiments. They found that agglomeration occurred in two stages. During the first stage, the nZVI particles rapidly agglomerated to form discrete micrometer-sized clusters. These clusters then linked to form chain-like fractal structures in the second stage. The rate of agglomeration depended on the particle concentration and was affected by the magnetic forces between particles, in addition to the forces discussed in Chapter 6. Agglomeration occurred rapidly: for a 2 milligram per liter (2 mg/L) solution of 20-nm nZVI particles, the first stage of agglomeration occurred in 10 min. These results illustrate why some nZVI reagents are modified, by the inclusion of polymers or other additives, to limit agglomeration.

10.1.4 IN SITU REMEDIATION WITH nZVI

Manufacturers typically ship nZVI reagents to a site in a concentrated slurry. It may be shipped at a high pH or under nitrogen atmosphere to limit passivation. Workers at the site dilute this slurry to the desired concentration. As described for two case studies in Section 10.1.6, this concentration is on the order of 2 grams per liter (g/L).

This diluted slurry can be injected into wells under pressure or by direct push installation. The term "direct push installation" refers to the technique of using hydraulic pressure to advance a tool string into the subsurface; this technique removes no soil and creates only a small borehole through which reagents can be injected.

Once injected, the fate and transport of nZVI depends not only on the characteristics of the reagent, but also on the flow of groundwater through the aquifer, the groundwater geochemistry, and the nature of the aquifer materials. nZVI can oxidize rapidly and agglomerate and attach to soil grains readily, reducing its reactivity and mobility [3, 5, 12, 14–16]. The mechanisms and rates of reaction are not yet well understood. Laboratory studies have found that the activity of nZVI particles depends on the particle type, pH, presence of compounds other than iron, amount of iron available in the particle core for reaction, oxide coating on the particle, and other aspects of geochemistry. Depending on these factors, the reactivity of nZVI lasts on the order of weeks to months. Field data are limited, as the technology has been commercially available only since 2003. Some reports from field applications suggest that nZVI may be reactive for months after injection.

nZVI particles tend to agglomerate and attach to soil grains, reducing their effective distribution through a plume of contamination [9, 10]. Attachment to soil grains, according to some estimates, would remove 99% of the nanoparticles within a travel distance between a few meters and a few tens of meters under typical groundwater conditions [3, 9]. Further transport might be possible under high-velocity conditions or in bedrock fractures.

10.1.5 POTENTIAL RISKS

This chapter opened with one authority's caution about the use of free nanomaterials in environmental applications. The paragraphs below describe initial data regarding the potential hazards of nZVI and discuss risk management positions taken regarding its use.

Laboratory studies provide some information on the potential toxicity of nZVI. In one *in vitro* experiment, central nervous system microglia cells exposed to nano iron at 2 to 30 mg/L exhibited oxidative stress response and assimilated nZVI into the cells. Weisner et al. [9] characterized these data as "preliminary results." Brunner et al. [17] studied the *in vitro* toxicity of nano Fe_2O_3. (Recall that Fe_2O_3 can be part of the surface coating of nZVI.) The tests used human (mesothelioma MSTO-211H) and rodent (3T3 fibroblast) cell lines. The researchers measured the effects on mean cell culture activity and DNA content after dosing cell cultures with particles at concentrations between 3.75 and 15 mg/L for a 6-day exposure period, and 7.5 to 30 mg/L for a 3-day exposure period. The control test of nano tricalcium phosphate did not show any effects. At concentrations up to 30 mg/L, nano Fe_2O_3 affected slow-growing 3T3 cells only slightly. Faster-growing MSTO cells showed a greater response. A dose as low as 3.75 mg/L had a significant effect on cell culture activity and DNA content, and a dose above 7.5 mg/L was lethal. Brunner et al. [17] concluded that the toxicity was approximately 40 times greater than would result from iron ions alone, and attributed that increase in toxicity to a nanoparticle-specific cytotoxic effect. They characterized these tests as screening tests, and recommended that further research be performed.

Ongoing laboratory studies will provide additional information. For example, Alvarez and Weisner [18] are studying the microbial impacts of engineered nanoparticles, including nZVI, at Rice University. This research is occurring from July 2005 to May 2008. Theodorakis et al. [18] are studying the acute and developmental toxicity of metal oxide nanoparticles, including Fe_2O_3, to fish and frogs. This project will conclude in September 2008. Elder et al. [19, 20] are studying iron-oxide nanoparticle-induced oxidative stress and inflammation using *in vitro* and *in vivo* tests.

Limited data are available from field work. In one pilot study [21], workers injected BNP into a fractured sandstone aquifer to treat TCE. The BNP slurry comprised 11.2 kg Fe-Pd BNP in 6050 L solution, or approximately 2 g/L. Initially, the concentration of TCE was 14 mg/L and the oxidation-reduction potential (ORP) was 75 millivolts (mV). Upon addition of the BNP, the ORP dropped to -290 to -590 mV, indicating a reducing environment, and the concentration of TCE decreased rapidly. Workers tested the effects on the microbial population and found that "the results of sampling the microbial community before and after injection indicated there were no significant trends due to the injection."

Finally, the Material Safety Data Sheet (MSDS) provides toxicity information to workers handling nZVI. MSDS sheets were obtained from three nZVI manufacturers:

1. Toda Americas, Inc., provided MSDSs for two nZVI products used in environmental remediation: RNIP-10DS [22] and RNIP-M2 [23]. Both MSDSs indicate that the material is nonflammable and stable, and list ACGIH Threshold Limit Values (TLVs) for iron of 5 milligrams per cubic meter (mg/m^3) based on Fe_2O_3. This value corresponds to the exposure limit for iron oxide dust and fume [24], rather than pertaining to nZVI *per se*. The RNIP-10DS contains elemental iron (10 to 20%), magnetite (Fe_3O_4) (15 to 5%), and water. It may cause irritation to eyes and the mucous membranes in the nose and throat. RNIP-M2 contains elemental iron (5 to 17%), magnetite (12 to 1%), water-soluble polymer (2 to 4%), and water. The material is a black liquid at pH $\sim$ 12. It may irritate the skin, eyes, and cause inflammation.

2. Princeton Nanotech, LLC, authored an MSDS for a nano iron slurry that PARS Environmental, Inc. markets as Nano-Fe [25]. The MSDS indicates that the material, a viscous liquid between pH 5.5 and 6.7, is stable and presents a low fire or reactivity hazard. It indicates a moderate acute health hazard to humans; potential health effects include eye irritation upon direct contact, skin irritation on prolonged or repeated contact, and potential harm if swallowed in large quantities, noting that the product has not been tested as a whole. Ecological information is noted as not available.

3. The MSDS for PolyMetallix™ Nanoscale Iron [26] describes the product as a stable black aqueous suspension at pH 7 to 9 containing 10 to 60% iron and 40 to 60% iron oxide ($FeO.Fe_2O_3.Fe_3O_4$). It notes the potential for irritation of eyes, skin, and the respiratory tract (upon inhalation). Cautions are based on iron oxide fume or dust.

The toxicity information on these MSDSs appears to be based on the characteristics of bulk iron or iron oxides, and other constituents or characteristics (e.g., pH) of the material.

Do the benefits of using this technology outweigh the risks? Gaps in the exposure pathway between the injection of nZVI and potential receptors mean that we cannot completely "connect the dots" to definitively determine a hazard:

- Because some nZVI products may be shipped as a slurry with pH ca. 12, risks can result from handling highly caustic materials. Workers can manage if not eliminate the risks from exposure to nZVI reagents using appropriate precautions in the field and personal protective equipment.

- nZVI tends to react and agglomerate readily, limiting — but not eliminating — the potential for nZVI to persist indefinitely and, for example, be inadvertently taken up in a drinking water supply. Modifications to nZVI reagents to increase their mobility and persistence in groundwater increase the potential for nZVI to move beyond a treatment zone.

- nZVI is used at a limited number of contaminated sites; and because the groundwater is contaminated, exposure to the groundwater should be limited.

- If exposure occurs, some studies have shown potential effects on human cells. Laboratory tests, as described above, have shown that glial cells can engulf nZVI, and nZVI can then stimulate oxidative stress. However, the human body may limit the transport of nanoparticles to the brain. Nanoparticles generally cannot cross a healthy blood-brain barrier. Some nanoparticles may be able to migrate to the brain via the olfactory nerves upon inhalation [27]. As described above, screening tests for Fe_2O_3 on a human cell line showed increased toxicity relative to iron ions, with a lethal dose at 7.5 mg/L. The authors cautioned, however, that the validity of *in vitro* results for *in vivo* situations is very limited and also recommended further research.

As with any conclusion drawn from preliminary data, this interpretation should be revisited as additional studies are performed.

Absent an ability to "connect the dots," some parties continue to use nZVI. The U.S. Environmental Protection Agency (EPA) sponsors research into and the use of nZVI at hazardous waste sites, as discussed for one case study below. Others are more cautious. In 2007, DuPont evaluated the possible risks of using nZVI in environmental remediation [28]. (See Chapter 11 for more information.) They concluded that "DuPont would not consider using this technology at a DuPont site until the end products of the reactions following injection, or following a spill, are determined and adequately assessed.... DuPont will monitor the status of this technology to review and update the decision as additional information becomes available." Specific concerns included:

- Possible fire hazard from nZVI dried slurry and any materials used to clean up a spill; the potential should be determined and an appropriate warning included in the MSDS.

- Unclear fate of nZVI after a spill dries. If spilled nZVI forms iron hydroxides and salts, then risk would be minimal. If the reaction produces nano-sized iron oxide particles, additional information would be needed on environmental fate and toxicology.
- Unknown sensitivity of human skin to nZVI (and some concern due to high pH of solution).
- Ultimate fate of injected nZVI unknown. Products likely to be soluble iron hydroxides and salts, which would present no long-term concerns. If the reaction produces nano-sized iron oxide particles, then additional information is needed on environmental fate.
- Insufficient nZVI, contact time, and/or untested reactions can result in incomplete contaminant destruction. "Careful design and testing of treatment systems is necessary to avoid these potential problems."

Following are brief descriptions of instances where those responsible for groundwater remediation have chosen to use nZVI.

10.1.6 Case Studies

Table 10.1 summarizes several case studies of the use of nZVI. Two projects are described below in more detail.

10.1.6.1 Nease Chemical Site

Formerly a chemical manufacturing plant, the Nease Chemical Site in Ohio is now on the National Priorities List of Superfund sites. Soil, sediment, and groundwater contain over 150 contaminants, primarily chlorinated compounds. In 2005, the U.S. EPA signed a Record of Decision that included treatment of groundwater in bedrock by nZVI. Subsequent work has included bench- and pilot-scale studies [29–31].

Volatile organic compounds (VOCs) contaminate groundwater in both overburden and bedrock aquifers. The overburden varies from silty sand to silty clay. Bedrock, comprising sandstone, is fractured and groundwater flow occurs primarily in fractures. Dense nonaqueous phase liquid (DNAPL) contaminates the bedrock and the concentration of total dissolved VOCs exceeds 100 mg/L. VOCs include tetrachloroethylene (or perchloroethylene, abbreviated PCE), trichloroethylene (TCE), cis-1,2-dichloroethylene (DCE), dichlorobenzene, and benzene.

The initial bench-scale test examined the following factors:

- Treatment of both chlorinated and nonchlorinated contaminants
- Form of nZVI, including four different materials (mechanically produced or chemically precipitated nZVI, with and without palladium catalyst)
- nZVI dosage ranging from 0.05 to 10 g/L
- Influence of site soils
- Generation of byproducts

TABLE 10.1

Summary of Selected Case Studies on nZVI [29-34]

Site	Target Compounds and Initial Concentrations	Effect of Treatment	nZVI Addition	Notes
Nease Chemical Site, OH	[TCE] ~ 70 mg/L [PCE] ~ 20 mg/L [DCE]DNAPL	After 4 weeks, [PCE] decreased 38–88% [TCE] decreased 30–70% [DCE] increased 0–100%	BNP – nZVI with Pd	Pilot-scale test; work ongoing.
NAES Lakehurst, NJ	[Σ VOCs] ~ 360 μg/L [TCE] ~ 56 g/L	After 6 months, [Σ VOCs] decreased 74% [TCE] decreased 79% [DCE] decreased 83%	BNP – nZVI with Pd	Did not achieve reducing conditions. Potentially deactivated nZVI due to mixing with oxygenated water. Decrease in contaminant concentrations may have resulted, in part, from dilution
NAS Jacksonville, FL	TCE PCE 1,1,1-TCA 1,2-DCE	"Significant" reduction in TCE; some increases in cis-1,2-DCE and 1,1,1-TCA	BNP – nZVI with Pd	Did not achieve reducing conditions. Potentially deactivated nZVI due to mixing with oxygenated water, or used insufficient iron
Trenton Switchyard, NJ	[Σ VOCs] up to ~1,600 μg/L; VOCs included 1,1-DCA, 1,1-DCE, 1,1,1-TCA, 1,2-DCA, TCE	Decreased total VOC concentrations by up to 90% within 24 weeks after injection	NanoFe Plus™ (nZVI with catalyst and support additive) injected in slurry up to 30 g/L	Treatment significantly reduced ORP and increased pH in most monitoring wells
Launch Complex 34, FL	TCE DNAPL	After 5 months, [TCE] decreased 57–100%	Emulsified nZVI	Longer-term reduction potentially due to biodegradation

Note: Abbreviations: BNP – bimetallic nanoparticle; DCA – dichloroethane; DCE – dichloroethylene; DNAPL – dense nonaqueous phase liquid; PCE – perchloroethylene (tetrachloroethylene); Pd – palladium; TCA – trichloroethane; TCE – trichloroethylene; VOCs – volatile organic compounds.

Researchers performed approximately 200 jar tests on groundwater samples containing total VOCs at over 100 mg/L, including approximately 80 mg/L PCE and 20 mg/L TCE.

The tests showed that bimetallic particles comprising nanoscale iron coated with about 1 wt% palladium were more effective than nZVI in the short term, effecting rapid reductions in concentrations of chlorinated VOCs at iron concentrations of 2 to 5 g/L. In one test, 2 g/L nZVI/Pd reduced the PCE concentration from approximately 70 mg/L to near detection limits in 2 weeks. nZVI without palladium showed only partial treatment within 2 weeks. Benzene was not effectively treated, and in fact, benzene was generated from the reduction of 1,2-dichlorobenzene. Site soils did not seem to affect treatment.

Work then proceeded with a pilot test to verify the initial results under field conditions, assess geochemical changes in the aquifer during treatment, and evaluate the transport of nZVI, thereby providing a basis for full-scale design. The pilot began with slug tests and tracer tests to provide information on how the groundwater flow could transport nZVI. Based on the results of the bench-scale tests, the design team planned to inject 2 gallons per minute (gpm) of a ~3000-gallon nZVI slurry containing 100 kg nZVI over 3 to 4 days. The reagent arrived at the site as a parent slurry and was diluted with water on site to prepare a solution containing 10 g/L nZVI. The parent slurry contained 20% powdered soy to act as an organic dispersant. Most batches contained 1% palladium; the last few injections did not. The target *in situ* concentration of nZVI was 2 g/L [80].

Presumably due to the heterogeneity of the aquifer materials, the field team could not achieve the planned injection rates. A total of 2665 gallons of nZVI slurry was injected at a rate of 0.15 to 1.54 gpm over a period of 22 days.

Initial test results were available as of this writing. Based on data from monitoring wells within 10 to 20 feet (ft) of the injection well, treatment reduced the concentrations of PCE by 38 to 88% and TCE by 30 to 70% within 4 weeks. The concentrations of breakdown products methane and ethane increased, as did the concentration of DCE. Measurements after 8 and 12 weeks indicated stable or increasing concentrations of the target contaminants, likely originating from an untreated source area up-gradient from the test area.

Plans for full-scale treatment, including additional means to treat benzene, are under development.

10.1.6.2   Naval Air Engineering Station, New Jersey

Chlorinated compounds contaminate groundwater in two areas of the Naval Air Engineering Station (NAES) in Ocean County, New Jersey. The U.S. Navy used BNP to treat the groundwater [29, 32], performing a bench-scale test in 2001, pilot work in 2003, and full-scale remediation in 2005 and 2006.

The NAES is underlain by a coastal plain aquifer, consisting of sand with some clay and gravel. The depth to the water table is approximately 15 ft. Groundwater contains PCE, TCE, 1,1-trichloroethane (TCA), and degradation products such as DCE and vinyl chloride (VC). Total VOC concentrations ranged up to 360 micrograms per liter (µg/L), including TCE at up to 56 µg/L. Much of this contamination existed 45 to 60 ft below the water table.

Initial testing showed that bimetallic particles containing palladium performed more effectively than nZVI without a catalyst. Full-scale treatment with nZVI/Pd BNP from PARS Environmental proceeded in two phases. Phase I, in November 2005, entailed injection of 2300 lb BNP. Workers injected a slurry containing 20 lb nZVI/Pd in 1200 gallons of water (or ~2 g/L) in each of 15 Geoprobe™ injection points. (Ten injection points were located in the northern plume, and five within the southern plume.) These injection points targeted the aquifer zone between 50 and 70 ft below ground surface (ft bgs) in 2-ft intervals.

The field team collected groundwater samples for 6 months after treatment. The concentrations of chlorinated compounds in some wells increased after 1 week, potentially due to desorption from soil. Concentrations subsequently decreased. The average decrease in the concentration of total VOCs in all monitoring wells was 74%; of TCE, 79%, and of DCE, 83%. ORP measurements indicated the general conditions in the aquifer. Six months after injection, ORP levels had decreased slightly in 3 of 13 monitoring wells, but increased or remained the same in other wells. These data showed that BNP injection did not create strong reducing conditions in the aquifer, possibly due to the oxygen in the water used to mix the BNP slurry at the site. pH levels were expected to rise significantly as a result of treatment; however, the average pH decreased slightly. Based on the geochemical data, the project team hypothesized that the decrease in VOC concentrations may have resulted from dilution. They inferred that mixing the nZVI slurry with a large volume of aerated water before injection passivated the nZVI [32].

Phase II occurred in January 2006. Workers injected a slurry containing 500 lb BNP using the same methodology as in Phase I. Monitoring continues as of mid-2007; groundwater quality standards have reportedly been achieved for some monitoring wells.

As the information in Section 10.1 shows, using nZVI has both benefits and possible risks. The next section discusses the development and use of other nanotechnologies in environmental remediation.

10.2   OTHER TECHNOLOGIES

Table 10.2 briefly describes technologies under development for wastewater treatment, environmental remediation, and related applications. It categorizes treatment technologies according to whether they rely on free nanoparticles or nanomaterials fixed in a matrix. This distinction may be important with respect to the potential for exposure to inadvertently released nanomaterials. Table 10.2 further categorizes treatment technologies according to their mode of treatment. Some technologies destroy contaminants by oxidation, reduction, or hydrolysis. Many such technologies incorporate nanocatalysts. Other technologies separate contaminants from groundwater or wastewater for further treatment or disposal.

Table 10.2 indicates the development status of each technology as of late 2007 — that is, bench scale, pilot scale, or full scale. Bench-scale tests are performed in

TABLE 10.2
Overview of Treatment Technologies based on Nanotechnology

Nanomaterial	Description	Compounds Treated	Development Status (2007)	Ref.
	Free Nanoparticles			
Treatment by degradation				
Nanoscale zero-valent iron (nZVI)	Slurry of nZVI injected into groundwater. Reduces some compounds (e.g., chlorinated solvents); promotes oxidation of others when oxygen is present in the aquifer.	Groundwater treatment: reduction of chlorinated ethylenes (trichloroethylene [TCE], tetrachloroethylene [PCE]), chlorinated ethanes, and PCBs. Oxidation of certain pesticides, phenol. Can adsorb As(III) and treat As(V) by reduction and adsorption.	Full-scale (groundwater treatment)	[3–5] [6–8] [21] [29–40]
	nZVI has been proposed as an amendment to sewage sludge (at < 0.1%).	Proposed use as biosolids amendment: complex with sulfur compounds, degrade toxic organics, and sequestrate Hg, Pb.		
Encapsulated ZVI	Variations of nZVI:		Bench scale	[40, 41]
• Silica nanoshells	Hollow silica shells containing ZVI are being studied for application to treatment of aqueous and nonaqueous liquids.	nZVI in silica nanoshell being evaluated to treat TCE.		
• Ferritin protein cage	Ferritin is an iron-storage protein with a nano ferrihydrite (iron oxide) core within a spherical protein cage. Researchers believe that this structure will allow control of the size and electronic structure of the particle to affect surface chemistry.	nZVI in ferritin protein cage used to treat sulfur dioxide; may have other catalytic applications.		

TABLE 10.2 (CONTINUED)
Overview of Treatment Technologies based on Nanotechnology

Nanomaterial	Description	Compounds Treated	Development Status (2007)	Ref.
Emulsified nZVI	nZVI injected into the subsurface in an emulsion comprising food-grade surfactant, biodegradable vegetable oil, and water. This formulation is thought to increase contact between dense nonaqueous phase liquid (DNAPL) contamination and nZVI; the vegetable oil component may also stimulate biological activity.	TCE	Full scale	[43]
Bimetallic nanoscale particles (BNP)	nZVI particles containing a trace coating of a catalyst, often palladium, are used to treat groundwater.	Reduction of chlorinated solvents (e.g., TCE, PCE), PCBs	Full scale	[32, 44]
Iron oxide nanoparticle catalyst for ozonation	Iron oxide (Fe_2O_3 and Fe_3O_4) catalyst for ozonation of organic compounds.	Parachlorobenzoic acid	Bench scale	[45]
Iron sulfide nanoparticles stabilized by biopolymers	Iron sulfide nanoparticles stabilized using a polymer from the basidiomycetous fungus, *Itajahia sp.* degrade chlorinated organic compounds.	Lindane (γ-hexachlorocyclohexane)	Bench scale	[46]
Microbially produced palladium nanocatalyst	Microbes recover palladium from wastewater from automotive catalyst or electronic scrap disposal, depositing Pd(0) on their cell surfaces. Bio-produced Pd nanocrystals catalyze the reduction of pollutants.	Cr(VI); PCBs	Bench scale	[47]
Nanostructured platinum-based catalysts	Nanostructured catalysts of Pt/TiO_2 and Pt/SiO_2 developed to catalyze destruction of NO_2 in automobile exhaust.	NO_2	Bench scale	[48]
Transition metal carbide nanoparticles as environmental nanocatalysts	Nanoparticles of molybdenum and tungsten carbides and oxycarbides used as an alternative to platinum catalysts to reduce NOx in gas stream.	NO_2	Bench scale	[49]

Name	Description	Application	Scale	Ref.
FAST-ACT	Reactive nanoparticles (unspecified composition) applied as a powder shaken onto surfaces or misted into the air. Various formulations for different applications.	Reportedly destroys nerve agents via hydrolysis and dehalogenation; neutralizes acids; absorbs spilled organic compounds.	Full scale	[50]
Nanocrystalline zeolite catalysts	Nano-scale zeolites (aluminosilicate molecular sieves) can catalyze the reduction of certain pollutants. Some zeolites can also adsorb certain contaminants.	Catalytic reduction of NO_2 in air. Adsorption of volatile organic compounds from air or water. Other applications.	Varied	[51, 52]
Treatment by separation				
Iron oxide (magnetite) nanoparticle sorbent	Iron oxide nanoparticles sorb contaminants from water, then are removed from water column via magnetic separation.	Arsenic, fluoride, phosphate, Cr(II), Cu(II), Ni(II)	Bench scale	[53–56]
Dendritic nanoscale chelating agents	Dendritic polymers (nanoparticles ca. 220 nm) chelate metal ions to enable their removal from water by ultrafiltration. Also proposed for use in removing metals from soils.	Metal ions (e.g., Cu(II))	Bench scale	[57, 58]
Functionalized titanium dioxide nanoparticles	Functionalized nanoparticles (FNP) created by coating 40–60 nm TiO_2 particles with an organosilane monolayer terminating with an ethylenediamene (EDA) ligand; FNPs treated with Cu(II) to create a $Cu(EDA)_2FNP$ that can bind certain anionic contaminants in water.	Pertechnetate (Tc-99)	Bench scale	[59]
Functionalized layered silicates	Thiol-functionalized silicate particles adsorb heavy metals.	Metal ions [Hg(II), Pb(II), Cd(II)]	Bench scale	[60]
Carbon nanotubes	Carbon nanotubes used to sorb contaminants, similar to activated carbon. (Note: Some of the relevant research has focused on the use of nanotubes as adsorbent materials in analytical chemistry, rather than in waste treatment *per se*.)	Cd, Mn, Ni, Pb, dichlorobenzene and other organic compounds	Bench scale	[61–66]

TABLE 10.2 (CONTINUED)
Overview of Treatment Technologies based on Nanotechnology

Nanomaterial	Description	Compounds Treated	Development Status (2007)	Ref.
Nanoscale biopolymers with tunable properties for improved decontamination and recycling of heavy metals	Nanoscale polymers produced by bacteria can adsorb heavy metals. Polymers can be "tuned" for selectivity.	Cd, Hg	Bench scale	[67]
Other forms of treatment				
Polyethylene glycol modified urethane acrylate (PMUA) nanoparticles	PMUA nanoparticles (~80 nm) have a hydrophilic exterior and hydrophilic interior (similar to a surfactant micelle). When added to sediments, these nanoparticles enhance the release of sorbed and sequestered hydrophobic organic contaminants (such as polynuclear aromatic hydrocarbons), making them more available for biodegradation.	Phenanthrene	Bench scale	[68]
Fixed Nanoparticles or Nanostructures				
Treatment by degradation				
Ferragels, comprising nZVI on a support medium, used to reduce target contaminants in water	"Ferragels" formed by reducing borohydride in the presence of a support material such as sand to create ZVI-coated particles that can be used in remediation. This approach was developed to avoid problems with agglomeration of nZVI.	Reduction and immobilization of Cr(V) and Pb(II)	Bench scale	[69]
Titanium dioxide catalysts	Used to catalyze photolysis of organic compounds.	Dyestuffs	Bench scale	[70]

	Description	Pollutant	Scale	Ref.
Titanium dioxide on activated fiber carbon cloth	Under ultraviolet light, used to photochemically degrade explosives residuals in water.	Hexahydro-1,3,5-trinitro-1,3,5-triazine (RDX) ad octahydro-1,3,5,7-tetranitro-1,3,5,7-tetraocine (HMX).	Bench scale	[71]
Titanium dioxide film on supporting substrates	Upon exposure to light (<40-nm wavelength), generates hydroxyl radical that oxidizes organic pollutants.	Organic pollutants including 17β-oestradiol, atrazine, formic acid. Also destroyed microorganisms: *E. coli*, *Clostridium perfringens*, and *Cryptosporum parvuum*.	Bench scale	[72]
Titanium dioxide in paints and window coatings	Designed as self-cleaning surfaces that destroy "dirt" when exposed to light. Experiments show that coatings can also destroy air pollutants, including NOx and toluene. Results for toluene dependent on humidity.	NO, toluene	Bench-scale demonstration; paints and coatings used commercially	[72]
Bimetallic palladium catalysts	Nanoparticles of bimetallic catalysts (palladium-alumina, palladium-gold) on solid support treat chlorinated compounds by reductive dechlorination.	TCE	Bench and pilot scale	[73]
Membrane-based nanostructured metals	Bimetallic catalysts (iron/nickel, iron/palladium) in ordered membrane domains to treat chlorinated compounds by reductive dechlorination.	TCE, PCBs	Bench scale	[74]
Nanocrystalline diamond	Boron-doped diamond (BDD) electrode, coated with nanocrystalline diamond, oxidizes organic contaminants.	Various organic compounds, microorganisms.	Bench scale	[72]
Treatment by separation				
Hydrated iron (III) oxide nanoparticles on ion exchange resin	Nanoparticles (20–100 nm) of iron oxide dispersed within a polymeric ion exchange resin remove contaminants from water.	Perchlorate, As	Bench scale	[75]
Composite biosorbent of nano Fe_3O_4/ *Sphaerotilus natans*	Biosorbent; used to adsorb heavy metals from wastewater.	Cu, Zn	Bench scale	[76]

TABLE 10.2 (CONTINUED)
Overview of Treatment Technologies based on Nanotechnology

Nanomaterial	Description	Compounds Treated	Development Status (2007)	Ref.
Nanostructured chitosan membranes	Electrospun nanofiber chitosan membranes under development to treat aqueous and gaseous streams via filtration, disinfection, and metal binding.	Metals, microbes, other pollutants	Bench scale	[77]
Self-assembled monolayers on mesoporous supports (SAMMS)	Active monolayer on ceramic support with mesopores (20–200 Å) sorbs contaminants. Different functional groups allow for selectivity. SAMMS are to be available as a particle (5–15 μm), an extrudate that can be fitted into ion exchange systems, and in an impregnated membrane system. Proposed for use in water treatment and to stabilize sludges or sediments.	Thiol-SAMMS: Hg, Ag, Au, Cu, Cd, Pb Chelate-SAMMS: Cu, Ni, Co, Zn Anion-SAMMS: chromate, arsenate HOPO-SAMMS: Am, Np, Pu, Th, U	Bench scale	[78, 79]

a laboratory. In its simplest form, a bench-scale test is designed to show whether a technology works in broad terms. More elaborate bench-scale tests provide information on the kinetics of a degradation reaction, and/or test the limits of the technology. A pilot test is larger scale and more elaborate than a bench-scale test. Pilot tests are generally used to evaluate materials-handling limitations, mass-transfer limitations, and cost. A pilot-scale test provides more accurate information on the performance of a technology than a bench-scale test. At full scale, a technology is commercially available and has been used in the field at one or more sites.

The development status of technologies listed in Table 10.2 ranges from initial concept testing in the laboratory to full-scale application. nZVI is by far the most tested and used technology at this time. For the treatment methods now being tested at bench scale, successful development to full-scale application will depend on their effectiveness, economics, risks, and impediments such as potential fouling from natural groundwater constituents.

REFERENCES

1. The Royal Academy of Engineering, the Royal Society. 2004. Nanoscience and Nanotechnologies: Opportunities and Uncertainties, Chapter 5. 29 July. http://www.royal-soc.ac.uk. (Accessed October 15, 2006)
2. Sellers, K. 1999. *Fundamentals of Hazardous Waste Site Remediation*. Boca Raton, FL: Lewis Publishers/CRC Press, p. 151–152.
3. Zhang, W. 2003. Nanoscale iron particles for environmental remediation: An overview. *J. Nanoparticle Res.,* 5:323–332.
4. Joo, S.H., A.J. Feitz, D.L. Sedlak, and T.D. Waite. 2005. Quantification of the oxidizing capacity of nanoparticulate zero-valent iron. *Environ. Sci. Technol.,* 39:1263–1268. Published online December 24, 2004.
5. Liu, Y. and G.V. Lowry. 2006. Effect of particle age (Fe^0 content) and solution pH on nZVI reactivity: H_2 evolution and TCE dechlorination. *Environ. Sci. Technol.,* published online August 30, 2006.
6. Tratnyk, P.G. and R.L. Johnson. 2006. Nanotechnologies for environmental cleanup. *Nanotoday,* 1(2):44–48.
7. Zhang, W.X. 2005. Nano-Scale Iron Particles: Synthesis, Characterization, and Applications. Meeting Summary: U.S. EPA Workshop on Nanotechnology for Site Remediation. Washington, D.C. 20–21 October. http://www.frtr.gov/nano.
8. Vance, D. 2005. Evaluation of the Control of Reactivity and Longevity of Nano-Scale Colloids by the Method of Colloid Manufacture. Meeting Summary: *U.S. EPA Workshop on Nanotechnology for Site Remediation*. Washington, D.C. 20-21 October. http://www.frtr.gov/nano.
9. Wiesner, M.R., G.V. Lowry, P. Alvarez, D. Dionysiou, and P. Biswas. 2006. Assessing the risks of manufactured nanomaterials. *Environ . Sci. Technol.,* 40(14):4336–4345.
10. Salch, N., K. Sirk, Y. Liu, et al. 2007. Surface modifications enhance nanoiron transport and NAPL targeting in saturated porous media. *Environ. Eng. Sci.,* 24(1):45–57.
11. Gavaskar, A., L. Tatar, and W. Condit. 2005. Cost and Performance Report: Nanoscale Zero-Valent Iron Technologies for Source Remediation. Prepared for Naval Facilities Engineering Command, Engineering Service Center, Port Hueneme, CA. 29 August. Available at http://www.clu-in.org.
12. Nurmi, J.T., P.G. Tratynek, V. Sarathy, et al. 2005. Characterization and properties of metallic iron nanoparticles: Spectroscopy, electrochemistry, and kinetics. *Environ. Sci. Technol.,* 39:1221–1230. Published online December 16, 2004.

13. Particle Size Distribution Study of Polymetallix™ nZVI. Undated. http://www. polymetallix.com/PDF/Particle%20Size%20Distribution%20Study%20of%20Polyme tallix%20Nanoscale%20Iron.PDF (accessed August 7, 2007).
14. Lowry, G. 2005. Nanoiron in the Subsurface: How Far Will it Go and How Does it Change? Meeting Summary: *U.S. EPA Workshop on Nanotechnology for Site Remediation*. Washington, D.C. 20–21 October. http://www.frtr.gov/nano.
15. Gill, H. 2005. *In Situ* Groundwater Treatment Using Nanoiron: A Case Study. Meeting Summary: *U.S. EPA Workshop on Nanotechnology for Site Remediation*. Washington, D.C. 20–21 October. http://www.frtr.gov/nano.
16. Liu, Y., S.A. Majetich, R.D. Tilton, D.S. Sholl, and G.V. Lowry. 2005. TCE dechlorination rates, pathways, and efficiency of nanoscale iron particles with different properties. *Environ . Sci. Technol.*, 39:1338–1345. Published online January 6, 2005.
17. Brunner, T.J., P. Wick, P. Manser, et al. 2006. *In vitro* cytotoxicity of oxide nanoparticles: comparison to asbestos, silica, and the effect of particle solubility. *Environ. Sci. Technol.*, 40:4374–4381. Published online March 11, 2006.
18. U.S. Environmental Protection Agency. 2007. Nanotechnology: Research Projects — Toxicity. http://es.epa.gov/ncer/nano/research/nano_tox.html. (Accessed August 2007)
19. Elder, A.C.P. and H. Yang. 2004. Iron Oxide Nanoparticle-Induced Oxidative Stress and Inflammation. Description. http://cfpub.epa.gov/ncer_abstracts/index.cfm/fuseaction/display.abstractDetail/abstract/7136/report/0. (Accessed October 21, 2007)
20. Elder, A.C.P. and H. Yang. 2006. 2006 Progress Report: Iron Oxide Nanoparticle-Induced Oxidative Stress and Inflammation. http://cfpub.epa.gov/ncer_abstracts/index.cfm/fuseaction/display.abstractDetail/abstract/7136/report/2006. (Accessed October 21, 2007).
21. Gheorghiu, F., M. Christian, R. Venkatakrishnan, and W. Zhang. 2005. *In Situ* Treatments using Nano-Scale Zero Valent Iron Implemented in North America and Europe. *U.S. EPA Workshop on Nanotechnology for Site Remediation*. Washington, D.C. 20–21 October. http://www.frtr.gov/nano.
22. Toda Kogyo Corporation. 2005. Material Safety Data Sheet: RNIP-10DS (Surface Stabilized Iron Slurry). 22 August.
23. Toda Kogyo Corporation. 2006. Material Safety Data Sheet: RNIP-M2 (Surface Stabilized Iron Slurry). 16 October.
24. U.S. Department of Health and Human Services. 1994. *NIOSH Pocket Guide to Chemical Hazards*. Washington, D.C.: U.S. Government Printing Office.
25. Princeton Nanotech, LLC. 2006. Material Safety Data Sheet: Nanoiron Slurry. 10 January.
26. Polyflon Company, A Crane Co. Company. 2004. Material Safety Data Sheet: Polyflon PolyMetallix™ Particles, Activated Metal Oxide #001. Original Date August 1, 2003; Revision Date February 25, 2004.
27. Borm, P.J.A., D. Robbins, S. Haubold, et al. 2006. The potential risks of nanomaterials: A review carried out for ECETOC. *Particle Fibre Toxicol.*, 3:11. DOI: 10.1.1186/1473-8977-3-11. Available at http://www.particleandfibretoxicology.com/content/3/1/11. (Accessed September 30, 2007).
28. DuPont. 2007. Nanomaterial Risk Assessment Worksheet: Zero Valent Nano Sized Iron Nanoparticles (nZVI) for Environmental Remediation. http://www.environmentaldefense.org/documents/6554_nZVI_Summary.pdf. (Accessed June 27, 2007).
29. U.S. Environmental Protection Agency. 2007. Nanotechnology — Superfund Site Remediation. Presented at *Session 4 of RISKeLearning: Nanotechnology — Applications and Implications for Superfund*. Presented by: Marti Otto, EPA OSRTI and Mary Logan, RPM, EPA Region 5. 19 April. http://www.clu-in.org/conf/tio/nano4/. (Accessed April 19, 2007)

30. U.S. Environmental Protection Agency. 2006. Technology Update #1: Nanotechnology — Nease Chemical Site, Columbiana County, Ohio. September. http://www.epa.gov/region5/sites/nease/. (Accessed July 8, 2007)
31. U.S. Environmental Protection Agency. 2007. Technology Update #2: Nanotechnology - Nease Chemical Site, Columbiana County, Ohio. June. http://www.epa.gov/region5/sites/nease/. (Accessed July 8, 2007)
32. Gavaskar, A., L. Tatar, and W. Condit. 2005. Cost and Performance Report: Nanoscale Zero-Valent Iron Technologies for Source Remediation. Prepared for Naval Facilities Engineering Command, Engineering Service Center, Port Hueneme, CA. 29 August. Available at http://www.clu-in.org.
33. Varadhi, S.N., H. Gill, L.J. Apolodo, K. Liao, R.A. Blackman, and W.K. Wittman. 2005. Full-Scale Nanoiron Injection for Treatment of Groundwater Contaminated with Chlorinated Hydrocarbons. Presented at the *Natural Gas Technologies 2005 Conference*, Orlando, FL. 1 February.
34. Henn, K.W. and D. Waddill. 2005. Implementation of a Nanoscale Zero Valent Iron Remediation Demonstration. *U.S. EPA Workshop on Nanotechnology for Site Remediation*. Washington, DC. 20-21 October. http://www.frtr.gov/nano.
35. Lowry, G.V. and K.M. Johnson. 2004. Congener-specific dechlorination of dissolved PCBs by microscale and nanoscale zerovalent iron in a water/methanol solution. *Environ. Sci. Technol.*, 38(19):5208–5216. Published online August 25, 2004.
36. Joo, S.H., A.J. Feitz, D.L. Sedlak, and T.D. Waite. 2005. Quantification of the oxidizing capacity of nanoparticulate zero-valent iron. *Environ. Sci. Technol.*, 39(5):1263–1268.
37. Kanel, S.R., B. Manning, L. Charlet, and H. Choi. 2005. Removal of arsenic(III) from groundwater by nanoscale zero-valent iron. *Environ. Sci. Technol.*, 39(5):1291–1298.
38. Kanel, S.R., J.-M. Greneche, and H. Choi. 2006. Arsenic(V) removal from groundwater using nano scale zero-valent iron as a colloidal reactive barrier material. *Environ. Sci. Technol.*, 40(6):2045–2050.
39. Song, H. and E.R. Carraway. 2005. Reduction of chlorinated ethanes by nanosized zero-valent iron: Kinetics, pathways, and effects of reaction conditions. *Environ. Sci. Technol.*, 39(16):6237–6245.
40. Li, X.-Q., D.G. Brown, and W.-X. Zhang. 2007. Stabilization of biosolids with nanoscale zero-valent iron (nZVI). *J. Nanoparticle Res.*, 9(2):233–243.
41. Lu, Y. and V.T. John. 2005. Novel Nanostructured Catalysts for Environmental Remediation of Chlorinated Compounds. Project Description. EPA Grant No. GR832374. http://cfpub.epa.gov/ncer_abstracts/index.cfm/fuseaction/display.abstractDetail/abstract/7562/report/. (Accessed October 15, 2006)
42. Strongin, D.R., T. Douglas, and M.A.A. Schoonen. 2006. A Bioengineering Approach to Nanoparticle Based Environmental Remediation — Final Report. http://cfpub.epa.gov/ncer_abstracts/index.cfm/fuseaction/display.abstractDetail/abstract/2370/report/F. (Accessed October 15, 2006)
43. Quinn, J., C. Geiger, C. Clausen, et al. 2005. Field demonstration of DNAPL dehalogenation using emulsified zero-valent iron. *Environ. Sci. Technol.*, 39(5):1309–1318.
44. He, F. and D. Zhao. 2005. Preparation and characterization of a new class of starch-stabilized bimetallic nanoparticles for degradation of chlorinated hydrocarbons in water. *Environ. Sci. Technol.*, 39(9):3314–3320.
45. Jung, H., H. Park, J. Kim, et al. 2007. Preparation of biotic and abiotic iron oxide nanoparticles (IOnPs) and their properties and applications in heterogeneous catalytic oxidation. *Environ. Sci. Technol.*, 41(13):4147–4747.
46. Paknikar, K.M., V. Nagpal, A.V. Pethkar, and J.M. Rajwade. 2005. Degradation of lindane from aqueous solutions using iron sulfide nanoparticles stabilized by biopolymers. *Sci. Technol. Adv. Mater.*, 6(3-4):370–374.

47. Schaefer, A. 2005. Microbes turn industrial waste into a nanocatalyst. *Environ. Sci. Technol.* Online News, 21 December. http://pubs.acs.org/subscribe/journals/esthag-w/2005/dec/tech/as_nanocatalyst.html.

48. Senkan, S. 2005. Nanostructured Catalytic Materials for NOx Reduction Using Combinatorial Methodologies. *Proceedings — Nanotechnology and the Environment: Applications and Implications, Progress Review Workshop III.* Arlington, VA. 26–28 October.

49. Shah, S. I. 2005. Synthesis, Characterization, and Catalytic Studies of Transition Metal Carbide Nanoparticles as Environmental Nanocatalysts. *Proceedings — Nanotechnology and the Environment: Applications and Implications, Progress Review Workshop III.* Arlington, VA. 26–28 October.

50. NanoScale Materials Inc. 2004. FAST-ACT First Applied Sorbent Treatment — Against Chemical Threats. Chemical Hazard Containment System and Neutralization System Technical Report.

51. Larsen, S. 2005. Development of Nanocrystalline Zeolite Materials as Environmental Catalysts. *Proceedings — Nanotechnology and the Environment: Applications and Implications, Progress Review Workshop III.* Arlington, VA. 26–28 October.

52. Ratner, M. and D. Ratner. 2003. *Nanotechnology: A Gentle Introduction to the Next Big Idea.* p. 90–91. Upper Saddle River, NJ: Pearson Education, Inc., Publishing as Prentice Hall PTR.

53. Yavuz, C.T., J.T. Mayo, W.W. Yu, et al. 2006. Low-field magnetic separation of monodisperse Fe_3O_4 nanocrystals. *Science,* 314:964–967.

54. Yean, S., L. Cong, C.T. Yavuz, et al. 2005. Effect of magnetite particle size on adsorption and desorption of arsenite and arsenate. *J. Mater. Res.,* 20(12):3255–3264.

55. Eskandarpour, A., K. Sassa, Y. Bando, et al. 2007. Creation of nanomagnetite aggregated iron oxide hydroxide for magnetic removal of fluoride and phosphate from wastewater. *ISIJ Int.,* 47:558.

56. Hu, J., G. Chen, I.M.C. Lo. 2006. Selective removal of heavy metals from industrial wastewater using maghemite nanoparticle: Performance and mechanisms. *J. Envir. Eng.,* 132(7):709–715.

57. Diallo, M.S., S. Christie, P. Swaminathan, J.H. Johnson, and N.A. Goddard. 2005. Dendrimer enhanced ultrafiltration. 1. Recovery of Cu(II) from aqueous solutions using PAMAM dendrimers with ethylene diamine core and terminal NH_2 groups. *Environ. Sci. Technol.,* 39:1366–1377.

58. Xu, Y. and D. Zhao. 2005. Removal of copper from contaminated soil by use of poly(amidoamine) dendrimers. *Environ. Sci. Technol.,* 39(7):2369–2375.

59. Mattigod, S.V., G.E. Fryxell, K. Alford, et al. 2005. Functionalized TiO_2 nanoparticles for use for *in situ* anion immobilization. *Environ. Sci. Technol.,* 3918):7306–7310.

60. Lagadic, I.L. 2006. One-Step Prepared Organosilicate Nanomaterials as Heavy Metal Ion Adsorbents. Presented at the *American Chemical Society Meeting & Exposition,* Atlanta, GA. 26–30 March. www.chemistry.org/portal/a/ContentMgmtService/xresources/ACS/Subportals/geoc/GEOC_Atlanta06.pdf.

61. Rangel-Mendez, J.R., G. Andrade-Espinosa, E. Muñoz-Sandoval, M. Terrones, and H. Terrones. 2006. Acid Activated Bamboo-Type Carbon Nanotubes and Cup-Stacked-Type Carbon Nanostructures as Adsorbent Materials: Cadmium Removal from Water. Poster presented at *2006 Annual Meeting AICHE,* San Francisco, CA. 12–17 November.

62. Li, Y.-H., S. Wang, J. Wei, et al. 2002. Lead adsorption on carbon nanotubes. *Chem. Phys. Lett.,* 357(3-4):263–266.

63. Li, Y.H., Y.M. Zhao, W.B. Hu, et al. 2007. Carbon nanotubes — the promising adsorbent in wastewater treatment. *Journal of Physics: Conference Series International Conference on Nanoscience and Technology (ICN&T 2006),* 61:698–702.

64. Liang, P., Y.L. Liu, J. Zeng, and H. Lu. 2004. Multiwalled carbon nanotubes as solid-phase extraction adsorbent for the preconcentration of trace metal ions and their determination by inductively coupled plasma atomic emission spectrometry. *J. Anal. At. Spectrom.*, 19:1489–1492.

65. Li, Q.-L., D.-X. Yuan, and Q.-M. Lin. 2004. Evaluation of multi-walled carbon nanotubes as an adsorbent for trapping volatile organic compounds from environmental samples. *J. Chromatogr. A,*. 1026(1-2):283–288.

66. Zhou, Q., J. Xiao, and W. Wang. 2006. Using multi-walled carbon nanotubes as solid phase extraction adsorbents to determine dichlorodiphenyltrichloroethane and its metabolites at trace level in water samples by high performance liquid chromatography with UV detection. *J. Chromatogr. A.*, 1125(2):152–158.

67. Lao, U.L., G. Prabhukumar, J. Kostal, M. Matsumoto, A. Mulchandani, and W. Chen. 2004. Nanoscale Biopolymers with Customizable Properties for Heavy Metal Remediation. Presented at *STAR Progress Review Workshop — Nanotechnology and the Environment II.* 18–20 August. www.es.epa.gov/ncer/publications/meetings/8-18-04/pdf/wilfred_chen_aug_20.pdf. (Accessed October 15, 2007).

68. Tungittiplakorn, W., C. Cohen, and L.W. Lion. 2005. Engineered polymeric nanoparticles for bioremediation of hydrophobic contaminants. *Environ. Sci. Technol.*, 39(5):1354–1358.

69. Ponder, S.M., J.G. Davab, and T.E. Mallouk. 2000. Remediation of Cr(VI) and Pb(II) aqueous solutions using supported, nanoscale zero-valent iron. *Environ. Sci. Technol.*, 34(12):2564–2569.

70. Wang, J., F.Y. Wen, Z.H. Zhang, et al. 2005. Degradation of dyestuff wastewater using visible light in the presence of a novel nano TiO_2 catalyst doped with upconversion luminescent agent. *J. Environ. Sci. (China)*, 17(5):727–730.

71. Liu, Z., Y. He, F. Li, and Y. Liu. 2006. Photocatalytic treatment of RDX wastewater with nano-sized titanium dioxide. *Environ. Sci. Pollut. Res. Int.*, 13(5):328–332.

72. Rickerby, D. and M. Morrison. 2007. Report from the Workshop on Nanotechnologies for Environmental Remediation, JRC Ispra. 16–17 April 2007. http://www.nanowerk.com/nanotechnology/reports/reportpdf/report101.pdf. (Accessed November 26, 2007)

73. Nutt, M.O., J.B. Hughes, and M.S. Wong. 2005. Designing Pd-on-Au bimetallic nanoparticle catalysts for trichloroethene hydrodechlorination. *Environ. Sci. Technol.*, 39(6):1346–1353.

74. Bhattacharyya, D., L.G. Bachas, D. Lewis, D. Meyer, S.M.C. Ritchie, and Y. Tee. 2006. Final Report: Membrane-Based Nanostructured Metals for Reductive Degradation of Hazardous Organics at Room Temperature. EPA Grant Number R829621. http://cfpub.epa.gov/ncer_abstracts/index.cfm/fuseaction/display.abstractDetail/abstract/2172/report/F. (Accessed October 16, 2006)

75. Cumbel, L. and A.K. Sengupta. 2006. Arsenic removal using polymer-supported hydrated iron(III) oxide nanoparticles: role of Donnan membrane effect. *Environ. Sci. Technol.*, 39(17):6508–6515. Published online May 27, 2006.

76. Guan, X.H., Y.C. Qin, L.W. Wang, R. Yin, M. Lu, and Y.J. Yang. 2007. Study on the disposal process for removing heavy metal ions from wastewater by composite biosorbent of nano Fe_3O_4/*Sphaerotilus natans. J. Environ. Sci. (China)*, 28(2):436–440.

77. Kit, K., P.M. Davidson, J. Weiss, and S. Zivanovic. Nanostructured Membranes for Filtration, Disinfection, and Remediation of Aqueous and Gaseous Systems. EPA Grant Number GR8322372. Project Description. http://cfpub.epa.gov/ncer_abstracts/index.cfm/fuseaction/display.abstract.Detail/abstract/7548/report/0. (Accessed October 15, 2006)

78. Pacific Northwest National Laboratory. Undated. SAMMS Home. http://samms.pnl.gov. (Accessed October 14, 2007)

79. Pacific Northwest National Laboratory — Operated by Battelle for the U.S. Department of Energy. Undated. SAMMS Technical Summary, Rev. 3. http://samms.pnl.gov. (Accessed October 14, 2007)

80. Logan, M. 2007. Personal communication with author. U.S. EPA Region V. 22 October.

81. Phenrat, T., N. Saleh, K. Sirk, R.D. Tilton, and G.V. Lowry. 2007. Aggregation and sedimentation of aqueous nanoscale zerovalent iron dispersions. *Environ. Sci. Technol.*, 41, 284–290.

11 Balancing the Risks and Rewards

Kathleen Sellers
ARCADIS U.S., Inc.

CONTENTS

Nanotechnologies offer broad promise to use raw materials and energy more efficiently. Some applications offer medical hope or environmental protection. These rewards, however, must be balanced against the potential risks from manufacturing, using, and disposing of products containing nanomaterials. This chapter discusses tools to evaluate the balance between potential risks and rewards, beginning with the concept of Life Cycle Analysis (LCA).

11.1 LIFE CYCLE ANALYSIS (LCA)

Life Cycle Analysis (LCA), an integral part of the ISO environmental management standards (ISO 14040), uses a mass and energy balance to determine the potential effects of product manufacture on human health and the environment. More formally [1],

> "LCA is a technique for assessing the environmental aspects and potential impacts associated with a product by:
>
> - Compiling an inventory of relevant inputs and outputs of a product system;
> - Evaluating the potential environmental impacts associated with those inputs and outputs;
> - Interpreting the results of the inventory analysis and impact assessment phases in relation to the objectives of the study.

249

LCA studies the environmental aspects and potential impacts throughout the product's life (i.e., cradle to grave) from raw materials acquisition through production, use and disposal. The general categories of environmental issues needing consideration include resource use, human health, and ecological consequences."

The formal process of LCA uses very specific information to quantify the consequences of a particular product's manufacture, use, and disposal. In the developing world of nanotechnology, such specific information can be difficult to ascertain. Many manufacturing processes are still in scale-up; often, and understandably, these processes are proprietary. Further, as discussed in previous chapters of this book, relatively little quantitative information is known about the potential releases of nanomaterials during the use and disposal of products based on nanotechnology, and the toxicity of those releases if they occur. Relatively few LCAs of nanotechnology have been published [2–14]. Focusing primarily on safety and environmental protection, several stakeholders have developed paradigms to evaluate the balance between the risks and benefits of nanotechnology.

11.2 ADAPTATIONS TO NANOTECHNOLOGY

Three approaches to evaluating nanotechnology are described below:

1. Screening approach developed at a workshop sponsored by The Pew Charitable Trusts, the Woodrow Wilson International Center for Scholars/Project on Emerging Nanotechnologies, and the European Commission [3]
2. The Nano Risk Framework developed by the Environmental Defense – DuPont Nano Partnership [4]
3. The XL Insurance Database Protocol, applied to nanotechnology by researchers at Rice University, Golder Associates, and XL Insurance [8]

The brief summaries that follow illustrate the general mass balance methodologies; critical features that characterize risks; and the uncertainties in evaluating risks from newly developed materials for which little information may be available. These approaches represent two different points of focus: the first two approaches focus on the nanomaterials themselves, and the third approach focuses on the processes used to manufacture the nanomaterials. Either or both of these focal points may be appropriate for balancing the risks and rewards of a particular nanotechnology, depending on the manufacturing process, materials used in that process, quantities of the nanomaterial used in a commercial product, and the potential for exposure (including whether nanomaterials are free or fixed). Of necessity, this chapter cannot present all the nuances of these models, and the reader is encouraged to consult the cited reference materials for more information.

11.2.1 SCREENING APPROACH

The 2006 workshop "Nanotechnology and Life Cycle Assessment: A Systems Approach to Nanotechnology and the Environment" brought together stakeholders from industry, government, academia, and nongovernmental organizations to talk about the life cycle analysis of nanomaterials [3]. Recognizing the limitations of applying rigorous LCA to nanotechnology, workshop participants developed an alternative approach. This five-step screening process combines elements of LCA, risk analysis, and scenario analysis:

1. Check for obvious harm. Consider compliance with health, safety, and environmental regulations using conventional analyses.
2. Perform a traditional LCA, excluding toxicity impact assessment. Instead, focus on potential impacts such as global climate change, eutrophication, etc. If the benefits appear to be substantial, then proceed; if not, stop product development.
3. Perform a thorough toxicity and risk assessment (RA) of the product. The assessment must consider possible exposures in each life-cycle stage.
4. Combine the results of Steps 2 (LCA) and 3 (RA) to determine overall impacts.
5. Perform a scenario analysis to extrapolate the results of Step 4 to large-scale usage (e.g., look at the implications of using a very small quantity of a nanomaterial in billions of products).

The authors of this approach acknowledge its current limitations: unavailability of proprietary information, limited hazard and exposure data, and lack of standard tools to combine LCA and RA (Step 4).

11.2.2 NANO RISK FRAMEWORK

Environmental Defense, a U.S.-based non-profit environmental advocacy group, and the multi-national chemical company DuPont collaborated to develop the Nano Risk Framework [4]. In the words of the developers,

"The purpose of this Framework is to define a systematic and disciplined process for identifying, managing, and reducing potential environmental, health, and safety risks of engineered nanomaterials across all stages of a product's 'life cycle' — its full life from initial sourcing through manufacture, use, disposal or recycling, and ultimate fate. The Framework offers guidance on the key questions an organization should consider in developing applications of nanomaterials, and on the information needed to make sound risk evaluations and risk-management decisions. The Framework allows users flexibility in making such decisions in the presence of knowledge gaps — through the application of reasonable assumptions and appropriate risk-management practices. Further, the Framework describes a system for guiding information generation and updating assumptions, decisions, and practices with new information as it becomes available. And the Framework offers guidance on how to communicate information and decisions to key stakeholders."

The Framework differs from LCA, as defined in Section 11.1, in that it focuses on potential environmental, health, and safety risks. It does not consider resource inputs.

The Nano Risk Framework comprises six steps, as described briefly below.

Step 1: Describe Material and Application. This step generates an overview of the physical and chemical properties of the material, sources and manufacturing processes, and possible uses. The overview includes existing materials that the nanomaterial may replace, and bulk counterparts of the nanomaterial.

Step 2: Profile Life Cycle(s). This step includes three components. Each relies on compiled "base set" data to define the characteristics and hazards of a nanomaterial. Where those data are not available, the Framework suggests using reasonable worst-case default values or assumptions. Analysts can replace those default values with actual data as they become available. This approach will provide an initially conservative estimate of risk that can be refined if appropriate.

a. *Profile Life Cycle Properties.* Develop base set data on physical and chemical properties of the nanomaterial, including property changes throughout the full product life cycle. (See Section 2.3.2.)

b. *Profile Life Cycle Hazards.* Characterize the potential hazards to human health, the environment, and safety from exposure to this material throughout its life cycle. In this step, analysts compile four base sets of data: health hazards, environmental hazards, environmental fate, and safety. Standard methods are not yet available to measure some of these base set parameters for nanomaterials. Base set data on health hazards include short-term toxicity, skin sensitization/irritation, skin penetration, genetic toxicity tests, and other data. Base set environmental hazard data include acute aquatic toxicology and terrestrial toxicology (i.e., earthworms and plants), and may include additional data if needed. Recommended base set data on the environmental fate of nanomaterials include physical-chemical properties, adsorption-desorption coefficients (soil or sludge), and nanomaterial aggregation or disaggregation in applicable exposure media. They also include data pertaining to persistence, characterizing biodegradability, photodegradability, hydrolysis, and bioaccumulation. Finally, base set safety hazard data include flammability, explosivity, incompatibility, reactivity, and corrosivity.

c. *Profile Life Cycle Exposure.* Quantify the potential for human and environmental exposures throughout the product life cycle. This definition is deceptively simple. The analyst must consider opportunities for direct contact or release to the environment at multiple stages: manufacture, processing, use, distribution/storage, and post-use disposal, reuse, or recycling.

Step 3: Evaluate Risks. The information collected in Step 1 and Step 2 is combined to estimate the risks to human health and the environment for each life cycle stage. Depending on the availability of base set data, the initial estimates may range from qualitative to quantitative. The analyst must determine gaps in the life cycle profiles and either generate data to fill the gaps or make reasonable worst-case assumptions.

Step 4: Assess Risk Management. For each life cycle stage, determine the actions needed to reduce and control risks from known and reasonably anticipated activities. These actions could include product modifications, engineering or management controls, protective equipment, or risk communication such as warning labels. The product developer might even decide to abandon the product.

Step 5: Decide, Document, and Act. At this stage, a review team critically analyzes the results to decide how to proceed. The team documents and communicates the results, and determines the course of action for refining or updating the conclusions.

Step 6: Review and Adapt. This step ensures that the risk characterization and risk management protocols continue to evolve as new information becomes available.

The authors of the Framework developed several case studies to test the Framework. Three of the case studies pertained to materials targeted in this book: nano titanium dioxide, zero-valent iron, and carbon nanotubes. Tables 11.1 through 11.3 summarize those case studies [5–7].

11.2.3 XL INSURANCE DATABASE PROTOCOL

The preceding adaptations of LCA focused on the nanomaterials themselves. In contrast, researchers at Rice University, Golder Associates, and XL Insurance focused on the materials and processes used to manufacture nanomaterials [8, 9]. Their risk analysis used the XL Insurance Database Protocol, which is used to calculate insurance premiums for the chemical industry, to examine the industrial fabrication of five nanomaterials. Those included three of the nanomaterials discussed at length in this book: single-walled carbon nanotubes, C60 fullerenes, and nano-titanium dioxide. The risk analysis entailed the following steps, as shown in Figure 11.1.

1. Identify process and materials:
   a. Determine synthesis methods, based on process currently used for commercial production or on processes likely to be scaled up for commercial production.
   b. Create block flow diagram showing inputs to and outputs from the manufacturing process, omitting energy use.
2. Characterize materials and processes:
   a. Collect and characterize data on material properties. Note that these data pertain to the raw materials used to manufacture the nanomaterials and the byproducts of fabrication; they do not pertain to the nanomaterials themselves. Critical data include toxicity, as expressed by LC50 and LD50, water solubility, log K_{ow}, flammability, and expected emissions. These initial data may trigger the need for additional information according to the protocol, so characterization of material properties is an iterative step. The protocol uses the collected data on material properties to rank substances by relative risk.
   b. Define manufacturing processes according to characteristics that determine risks, that is, temperature, pressure, and enthalpy. Then, for each point in the process and for each of the substances involved in the manufacturing process (except the nanomaterial), identify these characteristics: amount present, role in the process, physical phase at the temperature and pressure specified; and potential emissions. This step allows the model to calculate the probability of exposure from an in-process accident and from normal operations.

TABLE 11.1
Case Study Using the Nano Risk Framework: Titanium Dioxide [7]

Framework Step	Analysis
1. Describe Material and Application	DuPont™ Light Stabilizer 210 is a surface-treated form of TiO_2. The product absorbs and scatters ultraviolet (UV) light; addition of this product to a polymer protects the material from UV damage when exposed to sunlight. DuPont™ Light Stabilizer 210 will be transported to plastics producers in plastic bags, where it will be combined with other ingredients and mixed with molten polymer; it will comprise <3% of the end product.Potential applications include outdoor furniture, toys, and sheeting to protect greenhouses. Use of light stabilizers will extend the product life and thereby reduce the volume of plastics being landfilled.
2. Profile Lifecycle(s)	DuPont™ Light Stabilizer 210 is a white powder with particle sizes centered in the range of 130–140 nm. 10–20 wt% falls within the nano range (i.e., <100 nm). The particles are dense polyhedral TiO_2 crystals surface treated to control chemical reactions. The particles cannot be broken down by mechanical action, and their composition will not substantially change throughout the life cycle.
	Toxicity studies showed no significant difference between the effects of DuPont™ Light Stabilizer 210 and pigmentary TiO_2. Toxicity testing demonstrated low hazard to fish and invertebrates and indicated medium concern for algae, potentially due to the light-blocking effects.
	Titanium occurs naturally in the environment. No established analytical method can distinguish between the titanium in DuPont™ Light Stabilizer 210 and naturally occurring titanium.
	No accepted protocols for assessing the bioaccumulation potential of nanomaterials exist.
	Worker exposure should be low under normal operating conditions. Monitoring during production and handling indicated that airborne concentrations were below the acceptable exposure limit of 2 mg/m$_3$. If exposure limits were exceeded, workers were to don half-mask respirators with P100 filters.
	Exposure is expected to be low throughout the product life cycle because potential worker exposure is well-managed; due to the low production, use of engineering controls, and properties of the material, releases to the environment should be minimal; and the polymer end product should retain the DuPont™ Light Stabilizer 210 unless incinerated. Emissions from incineration should be low due to the low concentrations and emission controls on incinerators.
3. Evaluate Risks	Toxicity studies showed no significant difference between the effects of DuPont™ Light Stabilizer 210 and pigmentary TiO_2; both show low hazard. Further, exposure should be limited. Therefore, "there are no substantive risk issues associated with manufacture, processing, use or disposal of DuPont™ Light Stabilizer."
4. Assess Risk Management	Based on the conclusions of Step 3, few additional risk management measures were recommended. Those included personnel scheduling and monitoring during non-routine activities, and developing recycling procedures. Some additional toxicity testing was contemplated.

TABLE 11.1(CONTINUED)

Case Study Using the Nano Risk Framework: Titanium Dioxide [7]

Framework Step	Analysis
5. Decide, Document, and Act	The review team accepted the recommendations made in Step 4 and approved moving forward to product announcement and commercialization.
6. Review and Adapt	DuPont has scheduled reviews of DuPont™ Light Stabilizer 210 in 2009 and then every 4 years thereafter. "As needed" risk reviews will occur if triggered by a change in applications, new information on hazard, or higher than anticipated production.
Summary of Outcome	DuPont approved commercial introduction of the product.

3. Determine relative risk:
   a. Qualitative assessment. In this component of the risk assessment, analysts review information on the properties of each material that contribute to either exposure (based on emission estimates) or hazard (based on properties such as LC50 and LD50), and then rank each material as low, medium, or high for each of these properties. The aggregate ranking provides a qualitative assessment of risk.
   b. XL Insurance Database Methodology. The protocol estimates risk for three scenarios based on the manufacturing process, the materials involved, and their characteristics:
      i. Incident risk from accidental exposure resulting from a process accident.
      ii. Normal operations risk from routine emissions during manufacture.
      iii. Latent contamination from long-term operations and the site of manufacture.

The researchers used this protocol to estimate risks from manufacturing several nanomaterials. Tables 11.4 through 11.6 and Figure 11.2 summarize the analysis of the risks from manufacturing single-walled nano-titanium dioxide, carbon nanotubes, and C60 fullerenes [8, 9].

For perspective, the research team also used the protocol to evaluate the risks from the manufacture of six products in more longstanding, common use. Those products included wine, refined petroleum, and aspirin. Figure 11.2 illustrates the XL Insurance Database scores for selected nanomaterials and these other commercial products, and indicates which materials in the manufacturing process contributed most to the estimated risk.

The research team acknowledged that process information may be difficult to obtain. They also noted that manufacturers will likely refine production processes, to make them more efficient and perhaps to recycle or reuse some materials, as the manufacture of nanomaterials becomes more routine. Nonetheless, this model provides a useful measure of the industrial risks from the manufacture of nanomaterials.

TABLE 11.2

Case Study Using the Nano Risk Framework: Nano Zero-Valent Iron [6]

Framework Step	Analysis
1. Describe Material and Application	Nano zero-valent iron in nano-sized particles (nZVI) serves as a reagent to dechlorinate compounds such as tetrachloroethylene in groundwater. Vendors ship a highly concentrated slurry of nZVI to a contaminated site, where it is mixed with water and injected into an aquifer via small-diameter wells. DuPont did not produce or use nZVI at the time of the case study.
2. Profile Lifecycle(s)	nZVI slurries contain iron particles manufactured by one of several processes. The properties of the iron particles vary, depending on the manufacturer. Additives used to stabilize the nZVI slurries also vary with the manufacturer. Information on both the nZVI particles and the stabilizers is proprietary. Environmental health and safety data from suppliers varied in quality and completeness, and may have represented larger-sized "simple iron powder" rather than nZVI. Toxicological properties have not been thoroughly investigated. Warnings included the potential for skin or eye irritation upon contact, irritation of mucous membranes and upper respiratory tract if inhaled, and may have a laxative effect if swallowed.
	Effective use of nZVI to treat chlorinated compounds in groundwater requires adequate contact between nZVI and the contaminants; incomplete destruction could generate toxic partial degradation products. Spent iron typically precipitates as carbonate or sulfide minerals.
3. Evaluate Risks	The case study did not include a risk assessment due to the stage of the technology and DuPont's decision not to apply the technology.
4. Assess Risk Management	The case study did not evaluate risk mitigation measures due to the stage of the technology and DuPont's decision not to apply the technology.
5. Decide, Document, and Act	"DuPont would not consider using this technology at a DuPont site until the end products of the reactions following injection, or following a spill, are determined and adequately assessed." The case study identified five specific questions that must be addressed.
6. Review and Adapt	"DuPont will monitor the status of this technology to review and update the decision as additional information becomes available."
Summary of Outcome	Based on information available as of March 2007, DuPont has no immediate plans to implement this technology at any DuPont site.

TABLE 11.3

Case Study Using the Nano Risk Framework: Carbon Nanotubes [5]

Framework Step	Analysis
1. Describe Material and Application	DuPont considered incorporating carbon nanotubes (CNTs) into engineering thermoplastics to improve mechanical and electrical properties.
2. Profile Lifecycle(s)	Many of the CNT base set data were not available. DuPont purchased CNTs from outside suppliers in the form of powder (containing 96–100% CNTs) or encapsulated in polymer pellets (5–50 wt% CNTs). Absent clear environmental health and safety data, established exposure limits for CNTs, or toxicity data for the specific CNTs used, DuPont assumed CNTs were potentially hazardous. Air sample monitoring occurred during CNT handling and demonstrated the effectiveness of engineering controls.
	Because this was a research and development (R&D) project, the exposure analysis focused on workers rather than downstream users; such exposures would be considered if the products were to enter later stages of development.
3. Evaluate Risks	The evaluation did not include a systematic evaluation of risk because of the development stage.
4. Assess Risk Management	During R&D, DuPont chose to handle CNTs as hazardous material. Risk mitigation measures would be refined if nanocomposite products moved to full production.
5. Decide, Document, and Act	During R&D, personnel handled small quantities of CNTs in ways that minimized exposure, utilizing engineering controls, personal protective equipment, and special operating procedures. Air monitoring demonstrated the effectiveness of these measures.
6. Review and Adapt	The use of CNTs was under continuous review during the R&D process.
Summary of Outcome	Research project halted before commercialization for business reasons.

11.3 SUMMARY AND CONCLUSIONS

Development of alternative materials and new catalysts based on nanotechnology offers many potential benefits to human health and the environment. New technologies may save energy, use raw materials more efficiently, produce less waste, detect and treat environmental pollutants, and offer radically effective approaches to diagnosing and treating disease.

As with any new technological development, these benefits may come at some cost. Chapter 1 described the unintended consequences of some past technological advancements. LCA offers one tool to anticipate and avoid — or at least control — the adverse effects of developing nanotechnologies, particularly while regulators are wrestling with how to apply environmental, worker safety, and consumer protection regulations to nanotechnologies.

Research into potential risks is beginning to produce results. *In vitro* tests of certain nanomaterials have shown effects on mammalian cell lines, and some laboratory bioassays have demonstrated toxic effects. The most crucial hazards may result from

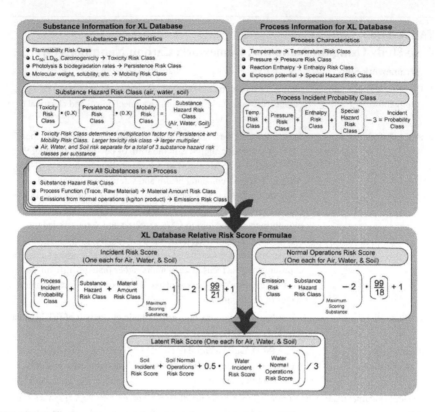

FIGURE 11.1 Schematic of the XL insurance database and formulation of risk scores [8]. (Reprinted with permission from Relative risk analysis of several manufactured nanomaterials: An insurance industry context. *Environ. Sci. Technol.*, 39(October):8985–8994. Copyright 2005, American Chemical Society.)

the inhalation of nanoparticulates, which can cause inflammation or immune-based response. While some laboratory results do give cause for concern, those concerns must be put into context. The methods of dosing test organisms may not reflect real-world conditions. Measures taken to prepare test solutions (for example, to keep nanomaterials in suspension) may introduce other toxicants or otherwise represent artificial conditions. In addition to the hazards presented by the nanomaterials themselves, one must consider the hazards posed by other materials used in the manufacturing process or part of the final product. Solutions of nZVI, for example, may be shipped at a highly caustic pH. Manufacture of C60 fullerenes, as another example, requires the use of highly toxic benzene.

For either a nanomaterial or an associated chemical to cause a risk requires a complete exposure pathway. That is, a mechanism must exist to transfer the compound or nanomaterial in question from the *source* in air, water, soil, sediment to the *receptor* organism in question. Exposure pathways may be complete during only portions of a product's lifecycle — during manufacture, perhaps, or during the use of a free (not fixed) nanoparticle. Little information is currently available on the end-of-life fate of nanomaterials used in commercial products or the potential for

TABLE 11.4

Case Study Using the XL Insurance Database: Nano Titanium Dioxide [8, 9]

Risk Analysis Step		Analysis
1. Identify process and materials		Hydrolysis and calcinations with chemical additives to control particle size; process currently in commercial use
2. Characterize materials and processes	A. Collect and characterize data on material properties	Data compiled for methane, hydrochloric acid, phosphoric acid, titanium tetrachloride, carbon dioxide.
	B. Define manufacturing processes, identify characteristics that determine risks	1. Prepare aqueous solution of $TiCl_4$ in solution with HCl, HPO_4. 2. Vacuum-dry solution and spray-dry at 200–250°C to produce dry TiO_2. 3. Calcinate at 600–900°C for 0.5–8 hours to produce crystalline nanostructure. 4. Wash precipitate with C_2H_5OH, dry, and mill to nano-sized particles.
3. Determine relative risk	A. Qualitative Assessment	Materials with very high risk include phosphoric acid (toxicity), titanium tetrachloride (toxicity).
	B. XL Insurance Database Methodology	

exposure. The tendency for many nanomaterials to agglomerate or sorb to solids may limit that potential.

In the end, the field of nanotechnology is too broad and as yet there are too many unknowns for gross generalizations regarding risks and rewards. Some applications offer true innovation and possible solutions to near-intractable problems; other nano promises may be largely marketing hype. Some hazards — specific to particular materials and exposures — may present significant risks that warrant careful control and monitoring. Others may fall within the range that society deems acceptable. At this stage in our understanding of nanotechnology and the environment, Albert Einstein may have offered the best advice: "Learn from yesterday, live for today, hope for tomorrow. The important thing is not to stop questioning."

TABLE 11.5

Case Study Using the XL Insurance Database: Single-Walled Carbon Nanotubes (SWNT) [8, 9]

Risk Analysis Step		Analysis
1. Identify process and materials		HiPco process of gas-phase chemical-vapor-deposition; process currently in commercial use.
2. Characterize materials and processes	A. Collect and characterize data on material properties	Data compiled for carbon monoxide, sodium hydroxide, iron pentacarbonyl, carbon dioxide, water.
	B. Define manufacturing processes, identify characteristics that determine risks	1. Introduce $Fe(CO)_5$ catalyst into injector flow via pressurized CO.
		2. Heat catalyst stream and mix with CO in graphite heater. $Fe(CO)_5$ decomposes to Fe clusters. Standard running conditions 450 psi CO pressure, 1050°C.
		3. C atoms coat and dissolve around the Fe clusters, forming nanotubes. Running conditions maintained 24–72 hours.
		4. Gas flow carries SWNTs and Fe particles out of the reactor. SWNTs condense on filters. CO passes through NaOH absorbtion beds to remove CO_2 and H_2O, then recycled.
3. Determine relative risk	A. Qualitative Assessment	No materials present very high risk according to this model. Materials with relatively high risk: carbon monoxide (emissions), iron pentacarbonyl (emissions) sodium hydroxide (toxicity, solubility), carbon dioxide (solubility, emissions).
	B. XL Insurance Database Methodology	See summary of results in Figure 11.2.

TABLE 11.6

Case Study Using the XL Insurance Database: Fullerenes [8, 9]

Risk Analysis Step		Analysis
1. Identify process and materials		Production in laminar benzene-oxygen argon flame; proprietary process modified from reference used for mass production.
2. Characterize materials and processes	A. Collect and characterize data on material properties	Data compiled for benzene, toluene, argon, nitrogen, oxygen, soot, activated carbon, carbon dioxide, water.
	B. Define manufacturing processes, identify characteristics that determine risks	1. Laminar flame of C_6H_6 and O_2, diluted with Ar. C:O ratio = 0.760. P = 12–100 torr. Flame operated 53–170 minutes.
		2. Sample of condensable compounds and soot taken via quartz probe.
		3. Sample weighed and extracted with C_7H_8, then filtered and concentrated by evaporation under N_2 stream.
		4. Concentrated solution of C60 and C70 in toluene separated on activated carbon.
		5. C60 filtrate concentrated with rotary evaporation and drying to 99% pure product.
3. Determine relative risk	A. Qualitative Assessment	Materials with very high risk include benzene (toxicity), soot (emissions).
	B. XL Insurance Database Methodology	See summary of results in Figure 11.2.

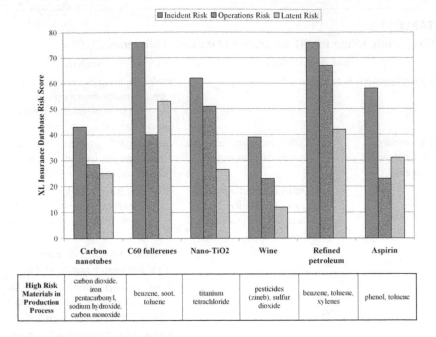

FIGURE 11.2 Manufacturing risks calculated using XL insurance database [8].

REFERENCES

1. International Organization for Standardization. 2006. Life Cycle Assessment — Principles and Framework. ISO 14040.
2. Sweet, L. and B. Strohm. 2006. Nanotechnology — life-cycle risk management. *Hum. Ecolog. Risk Asses.*, 12:528–551.
3. Klöpffer, W., M.A. Curran, P. Frankl, R. Heijungs, A. Köhler, and S.I. Olsen. 2006. Nanotechnology and Life Cycle Assessment: a Systems Approach to Nanotechnology and the Environment. Synthesis of Results Obtained at a Workshop, Washington, D.C. 2–3 October. Organized by the European Commission and the Woodrow Wilson International Center for Scholars, Project on Emerging Nanotechnologies. 20 March 2007. www.nanotechproject.org. (Accessed May 1, 2007)
4. Environmental Defense and DuPont. 2007. Nano Risk Framework. 21 June. http://www.NanoRiskFramework.com. (Accessed June 27, 2007)
5. Environmental Defense and DuPont. 2007. Nanomaterial Risk Assessment Worksheet: DuPont™ Light Stabilizer for use as a polymer additive. http://www.environmentaldefense.org/documents/6552_TiO2_Summary.pdf. (Accessed June 27, 2007)
6. Environmental Defense and DuPont. 2007. Nanomaterial Risk Assessment Worksheet: Zero Valent Nano Sized Iron Nanoparticles (nZVI) for Environmental Remediation. http://www.environmentaldefense.org/documents/6554_nZVI_Summary.pdf. (Accessed June 27, 2007)
7. Environmental Defense and DuPont. 2007. Nanomaterial Risk Assessment Worksheet: Incorporation of Single and Multi Walled Carbon Nano Tubes (CNTs) into Polymer Nanocomposites by Melt Processing. http://www.environmentaldefense.org/documents/6553_CNTs_Summary.pdf. (Accessed June 27, 2007)

8. Robichaud, C.O., D. Tanzil, U. Weilenmann, and M.R. Wiesner. 2005. Relative risk analysis of several manufactured nanomaterials: An insurance industry context. *Environ. Sci. Technol.*, 39 (October):8985–8994.

9. Robichaud, C.O., D. Tanzil, U. Weilenmann, and M.R. Wiesner. 2005. Supporting information for relative risk analysis of several manufactured nanomaterials: An insurance industry context. *Environ. Sci. Technol.*, published online October 4, 2005. http://pubs.acs.org/subscribe/journals/esthag/suppinfo/es0506509/es0506509si20050812_025756.pdf. (Accessed July 4, 2007)

10. Olsen, S.I. 2005. Life Cycle Assessment of Micro/Nano Products. Presented at *Nanotechnology and OSWER: New Opportunities and Challenges*, Washington, D.C. July 12–13, 2006. http://www.epa.gov/oswer/nanotechnology/events/OSWER2006/pdfs/.

11. Lloyd, S.M. and L.B. Lave. 2003. Life cycle economic and environmental implications of using nanocomposites in automobiles. *Environ. Sci. Technol.*, 37(15):3458–3466.

12. Lloyd S.M., L.B. Lave, and H.S. Matthews. 2005. Life cycle benefits of using nanotechnology to stabilize platinum-group metal particles in automotive catalysts. *Environ. Sci. Technol.*, 39(5):1384–1392.

13. Bakshi, B.R. 2005. Evaluating the Impacts of Nanotechnology via Thermodynamic and Life-Cycle Analysis. *Proceedings — Nanotechnology and the Environment: Applications and Implications, Progress Review Workshop III*. Arlington, VA. 26–28 October. http://es.epa.gov/ncer/publications/workshop/pdf/10_26_05proceeding1.pdf.

14. Khannaa, V., B.R. Bakshi, and L.J. Lee. 2007. Life Cycle Energy Analysis and Environmental Life Cycle Assessment of Carbon Nanofibers Production. *Proceedings — 2007 IEEE International Symposium on Electronics & the Environment*. Orlando, FL. 7–10 May.

Index

A

S